# Assessment of Drug Interactions & Drugs Resulting in QT Interval Prolongation

First Printing: 2019

ISBN: 978-1-79482-893-3

**Affiliation of Author**

Afzoon Butool,
Guru Nanak Institutions Technical Campus-School of Pharmacy,
Ibarhimpatnam, Ranga Reddy District, Hyderabad,
Telangana, India- 501506.


N Sushrusha,
Guru Nanak Institutions Technical Campus-School of Pharmacy,
Ibarhimpatnam, Ranga Reddy District, Hyderabad,
Telangana, India- 501506.


Sagar Pamu,
Guru Nanak Institutions Technical Campus-School of Pharmacy,
Ibarhimpatnam, Ranga Reddy District, Hyderabad,
Telangana, India- 501506.

www.lulu.com
Lulu Press, Inc
627 Davis Drive, Suite 300, Morrisville, NC 27560.

# Assessment of Drug Interactions & Drugs Resulting in QT Interval Prolongation

**Author**

Afzoon Butool

N Sushrusha

Sagar Pamu

**Editor**

Sagar Pamu

2019

## About Author

**Afzoon Butool** pursued Pharm. D in Guru Nanak Institutions Technical Campus-School of Pharmacy. She excelled topper in academics and passionate in research. She has an experience in patient counselling, ADR detection & reporting, identifying interventions in patient's prescription during clerkship, internship and health camps. She also participated in various conferences and seminars.

**Shushrusha Nall**a pursued Pharm. D in Guru Nanak Institutions Technical Campus-School of Pharmacy. She has participated in various conferences and seminars. She has good skills in ADR detection and prescription analysis.

**Sagar Pamu**, Pharm. D, is an Assistant Professor in Guru Nanak Institutions Technical Campus- School of Pharmacy. He filed one patent, authored 19 articles in international & national journals and 19 books. He also participated and presented in various national conferences and seminars.

# Acknowledgement

At the very outset, we thank God, the Almighty for showering his blessings and being a source of guidance and wisdom throughout the study without which no human achievement is possible.

We are indebted to our beloved **Parents** without whose encouragement and help our professional career would never seen the light of the day.

As we walk along the path of life, we have the pleasure of meeting people who search our life in such a way that it never is the same gain…it may be small thoughtful things they do, a smile, a helping hand, a word of encouragement or just by mere presence they make our life worth living.

Accomplishing this project has been a great learning and very fulfilling experience. There have been many people who have come along side and helped in conceiving, designing, and executing this project. I would like to place a record and my sincere appreciation for their contribution.

"Words cannot be said nor written for obligation and indebtedness"

We take immense pleasure in thanking our research guide **Dr. C. Narasimhan, MD, DM, AB (USA), Consultant cardiologist and electrophysiologist, Head Of Department of cardiology, Care Hospitals** for his rejuvenating inspiration, kind co-operation, valuable guidance, suggestions and encouragement throughout the progress of this work which helped us to complete every aspect of this project work and for his generous offer to use the available data on QT prolongation patients.

We wish to extend our sincere thanks to our co-guides **Dr. Daljeet Kaur Saggu, MD, DM, Consultant cardiologist and electrophysiologist**, **Dr. Sachin Yalagudri,Consultant Cardiologist** for their kind co-operation, valuable guidance, suggestions and encouragement at every step, and continuous assistance right from conceptualization of the project work for the preparation of this thesis.We also thank our fellow sir **Dr. Vihang Shah, DNB** for his valuable guidance during our project.

We heartily thank **Dr. P. Sagar, Pharm. D (Ph. D), Asisstant professor,Department of PharmacyPractice,** our research guide for his keen interest, valuable guidance, logical thoughts, kind co-operation, constructive criticism, and encouragement in every step, and continuous assistance right from conceptualization of the project work to the preparation of this thesis and suggestions which helped us to complete every aspect of this project work.

We are deeply grateful to **Dr T. Lakshmi , Head of the Department, Pharm. D** for her guidance, patience and support. We consider our self very fortunate to being able to work with a very considerate and encouraging professor like her.

Just as a music conductor is important for an orchestra, the Principal of a college is pivotal. Our principal, teacher **Dr. T. Rama Rao, Associate Director and Principal,** never went back in taking personal interest, providing constant encouragement and valuable advice. I am deeply thankful for facilitating the project and creating a conductive environment for completing the project.e are immensely thankful to **Dr. B Soma Raju, Chairman and Managing Director, Care Hospitals**, for his rejuvenating inspiration, valuable suggestions and encouragement given to us during the project work.

We would like to express our deep sense of gratitude to **Dr. N Krishna Reddy, Vice Chairman,** for his guidance, valuable suggestions and affection during our course.

We are immensely thankful to **Dr. Raghava Raju, Medical Director**, for his kind co-operation, inspiring guidance, supervision and help in completing this work.

We would like to express our sincere gratitude and feel immense pleasure in thanking to **Dr. Wali mohammad, Pharm. D, Clinical research department, Dr. Nalla Swapna,Physician assistant and Dr. Man Mohan Lal, Physician Assistant** for their guidance during the term of our project. Without their valuable assistance, this work would not have been completed.

We take the privilege to express our heart full gratitude to **Dr. Gopi Krishna, Medical Superintendent,** for his co-operation, affection, encouragement and moral support throughout our project.

We express our sincere thanks to **Mr. B Venkatesh, Clinical Pharmacist and Research Coordinate, and Mr. Vidya Sagar, Clinical Pharmacist** for their cooperation and help in completing this work.

Our sincere thanks to **Mr. Sudheer** for his help and guidance in performing statistics of the study.

I am thankful to **Teaching and Non- Teaching staff of Guru Nanak Institutions (Pharmacy)** for their help and encouragement during the course of the study.

Finally, yet importantly, we thank all the **Patients** who participated in the study without whom the study would not be possible.

Thanks again.......

Afzoon Butool, N Sushrusha

Dedicated to my beloved Parents, Teachers and Friends

Thank you. Without your support and patience, I would have never achieved my dream

# Abbreviations

| Abbreviated Form | Full Form |
|---|---|
| ADHF | Acute Decompensated Heart Failure |
| ADR | Adverse Drug Reaction |
| AF | Atrial Fibrillation |
| AV | Atria Ventricular |
| CHF | Congestive Heart Failure |
| CLQT | Congenital Long QT |
| CPMP | Committee For Proprietary Medicinal Products |
| CYP | Cytochrome P |
| DIPS | Drug Interaction Probability Score |
| DUR | Drug Utilisation Review |
| ECG | Electrocardiogram |
| HCM | Hypertrophic Cardio Myopathy |
| HERG | Human Ether -A-Go-Go- Related Gene |
| HR | Heart Rate |
| ICD | Implantable Cardio Vector Defibrillator |
| IKR | Inward Rectifier Current (Rapid) |
| IKS | Delayed Rectifier Current (Slow) |
| IV | Intravenous |
| LQTS | Long QT Syndrome |
| LV | Left Ventricular |
| M cells | Myocardial Cells |
| PAR | Population Attributable Risk |
| SNRI | Serotonin Norepinephrine Reuptake Inhibitors |
| SSRI | Selective Serotonin Reuptake Inhibitors |
| TDP | Torsades De Pointes |
| UMC | Uppsala Monitoring Centre |
| VF | Ventricular Fibrillation |
| VT | Ventricular Tachyarrhythmia |
| WHO | World Health Organisation |

# Table of Contents

## List of Tables

# List of Figures

# Chapter-1

# Introduction

On electrocardiogram (ECG) QT interval is estimated from the beginning of Q wave to the end of T wave on the horizontal axis which is shown in figure 1.1. Its measurement constitutes the time of QRS complex, the extent of ST-segment and the width of T wave[1]. It characterizes the total duration of ventricular electrical activity I.e. the time it takes for the heart ventricles to depolarize and repolarize[2].

## ECG Recording of a Healthy Heartbeat

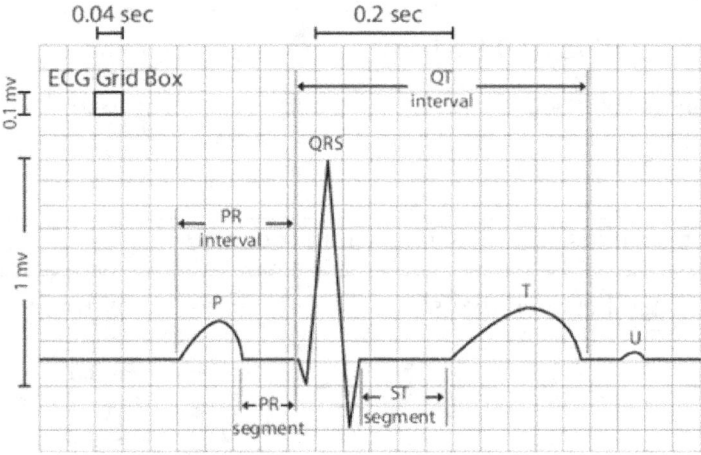

Figure No- 1.1: ECG Wave Form

## 1.1 Measurement of QT Interval:

The 12-lead ECG is the comprehensive primary diagnostic test in the estimation of the prolonged QT interval and is subject to mishandling, misconception and underestimation. The computerised calculation is agitated with errors, particularly in ECG's with complicated T-wave and U-waves with prominent dips and notches[3,4]. Therefore, physicians must be educated regarding the computerised QTc and they must understand that it cannot be trusted upon during the diagnosis of the long QT interval and should be calculated manually[5]. Usually, the long QT interval is assessed when QTc is higher than 440ms in men and higher than 460 ms in women[6].

The normal duration of the QT interval should be in the range of 0.35 to 0.43 seconds[1]. The normal, borderline and prolonged QT interval values are listed in table 1.1. The QT interval is in contrary to the heart rate I.e. it shortens during tachycardia and lengthens during bradycardia. This can be explained due to the shortening or lengthening of the ventricular repolarization. Thus, the QT interval needs to be corrected for what it would conceptually be at a heart rate of 60 beats per minute[7]. There are various formulae derived for the calculation of the QT interval:

1. Bazett formula =QT/RR1/2
2. Fridericia formula =QT/RR1/3
3. Framingham formula =QT+0.154 (1-RR)
4. Hodges formula =QT+0.00175 ([60/RR] 60)[9].

The most persistently used formula clinically was proposed by Bazzet. Bazzet's formula states that

QTc=QT interval in seconds/Root RR interval in seconds.

The RR interval is estimated between two successive R waves[7], shown in the figure: 1.2.

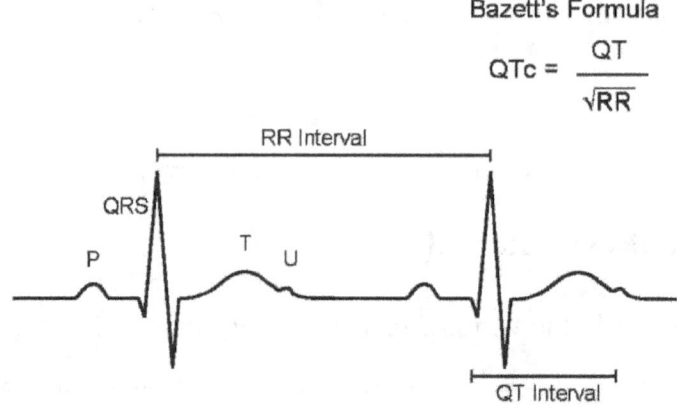

**Figure No- 1.2: ECG Showing RR Interval and QT Interval for Calculating Bazzets Formula**

QTc is estimated in at least three to four cardiac cycles in lead II or V5 and V6 using the tangent method with the longest value used[9]. The end of T wave is not always clear, it may be superimposed by U wave. The tangent method is the desirable method to define the end of T wave which is shown in figure 1.3. It is done by drawing the tangent on the maximum downslope of T wave which is intersected with a line drawn on isoelectric baseline[7].

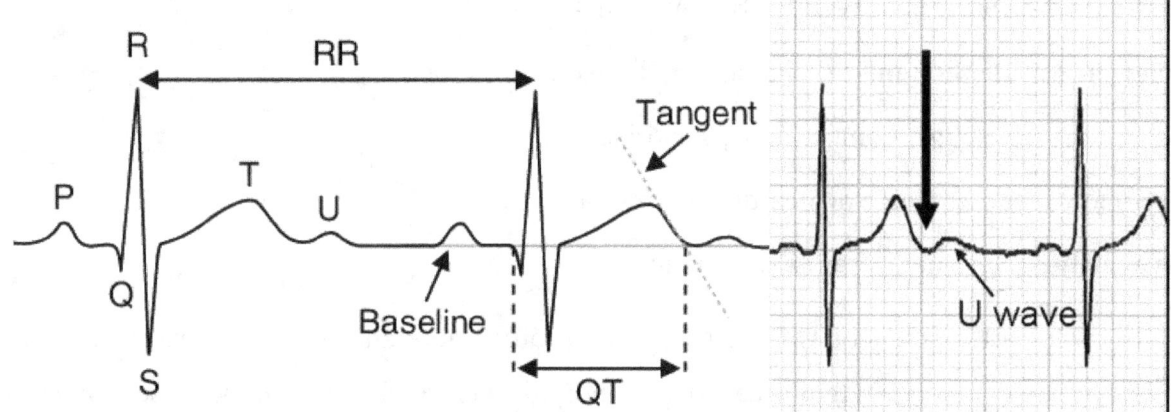

**Figure No- 1.3: ECG Showing QT Interval and U Wave**

**Table No-1.1: Table showing normal, borderline and prolonged QTc values[8]**

| Category | 1-15 years (ms) | Adult males (ms) | Adult females(ms) |
|---|---|---|---|
| Normal | <440 | <430 | <450 |
| Borderline | 440-460 | 430-450 | 450-470 |
| Prolonged | >460 | >450 | >470 |

Long QT interval can be congenital or drug-induced and is responsible for Torsades de pointes (TdP)[10]. It is susceptible to malignant ventricular tachyarrhythmias (VT), torsades de pointes, ventricular fibrillation and sudden cardiac death by prolonging repolarization, causing early after depolarizations and refractoriness[10,11].

## 1.2 Causes of QT Interval Prolongation:

❖ **Congenital:** Congenital long QT (CLQT) is an inherited cardiac disease distinguished by a QT interval prolongation at baseline ECG. Genetic mutations of the following genes precipitate the disease (*KCNQ1, SCN4B, CAV3, KCNE2, KCNH2, KCNE1, CAV3, SCN4B, CACNA1c*) by delaying the action potential of ventricular myocardial cells. The widespread congenital Long QT is the (Long QT syndrome) LQTS1 that is caused by a mutation in gene KCNQ1[12].

❖ **Acquired:** Acquired long QT interval can be due to the following[13]:

**Electrolyte insufficiency:** hypokalaemia, hypocalcaemia, hypomagnesemia

**Antiarrhythmic drugs:** sotalol, flecainide, amiodarone

**Bradyarrhythmia:** sinus bradycardia, AV block

**Tricyclic antidepressants:** escitalopram, duloxetine

**Miscellaneous drugs:** domperidone, cisapride

**Coronary disease:** Acute myocardial infarction

**Myocarditis:** viral myocarditis, rheumatic fever.

The prolongation of QT interval is clinically important because it can lead to the typical type of new arrhythmia called polymorphic ventricular arrhythmia also known as Torsades de Pointes which precisely means torsion around a point[14]. This means that the structure and morphology of the polymorphic QRS complex keep changing in direction and amplitude[14]. This arrhythmia may stop immediately or deteriorate into ventricular fibrillation which is shown in figure1.4. It may also cause haemodynamic insult and can also lead to death[11].

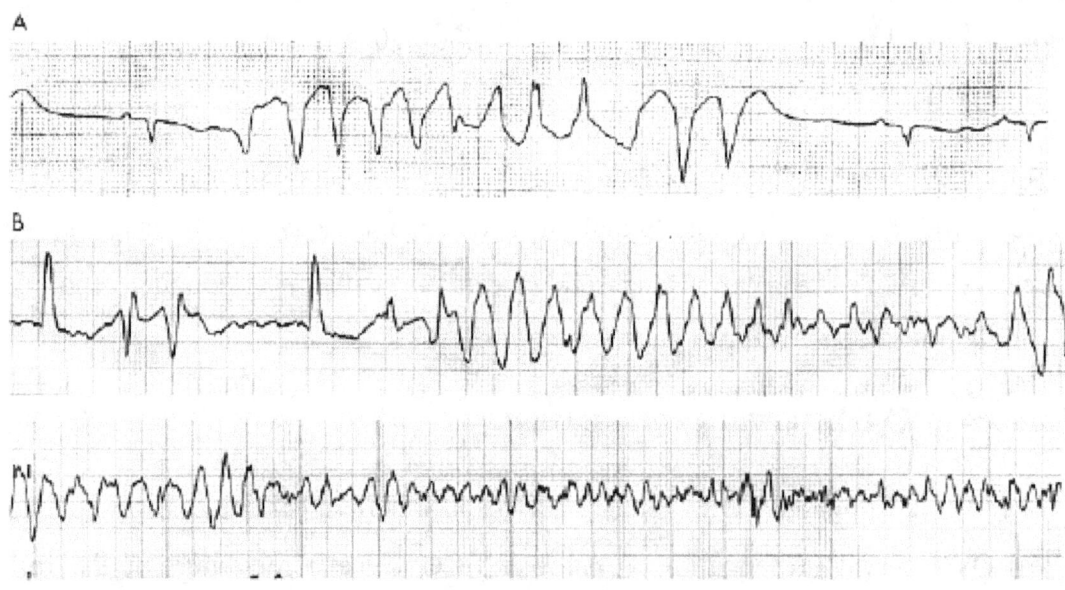

**Figure 1.4 Torsades de Pointes**

**(A) Self-limiting torsades de pointes**

**(B) TdP leading to ventricular fibrillation**

## 1.3 Clinical Presentation:

The symptoms of prolonged QT interval patients are variable, most of the patients are asymptomatic with QT interval prolongation on ECG, a very few are symptomatic and are present with palpitations, syncope or sudden cardiac death[15].

## 1.4 Risk factors for drug-induced QT prolongation/TdP [16] are listed below

in the table no: 1.2. Anyone who takes a QT-prolonging drug has some risk, but the higher risk is due to other factors that include the following.

Table No-1.2: Risk Factors of QT Interval Prolongation

| Potential risk factors: (modifiable) [17,18] | Unmodifiable risk factors: [19,20] |
|---|---|
| 1. Bradycardia, which includes recent conversion from Atrial Fibrillation (AF). | 1. Family history. |
| 2. Drugs that cause electrolyte abnormalities and drugs which can cause hepatic and renal impairment. | 2. Congenital long QT Syndrome [19]. |
| 3. Hypokalaemia  Hypomagnesemia  Hypocalcaemia  Hypothyroidism | 3. Female predisposition (70%). |
| 4. Drug interactions involving more than one QT-prolonging drug. | 4. Structural heart disease-Hypertrophic cardiomyopathy (HCM), Congestive heart failure (CHF), Left ventricular dysfunction. |
| 5. Drugs that inhibit the metabolism of other QT-prolonging drugs. | 5. Ageing |
| 6. Obesity and starvation [21]. | 6. Hepatic or renal impairment. |
| 7.Overdose or rapid IV (intravenous) administration. | 7. A family history of sudden death. |

## 1.5 Drugs that cause QT prolongation/TdP: The drugs which cause QT interval

prolongation or TdP are mentioned in table 3 and 4 and are vast [17]. Non-cardiac drugs that cause QT interval prolongation are listed in table no: 1.3 below [17]:

Table No-1.3: List of Non-cardiac Drugs Causing Long QT interval

| Antimicrobials | Antihistamines |
|---|---|
| Clarithromycin | Terfenadine |
| Chloroquine | Diphenhydramine |
| Moxifloxacin | Loratadine |
| Erythromycin | Astemizole |
| Levofloxacin | Mesolastine |
| Grepafloxacin | **Antiemetics/Gastric Motility agents** |
| **Antimalarial Drugs** | Cisapride |
| Halofantrine | Domperidone |
| Amantadine | Serotonin antagonists/agonists |
| **Immunosuppressants** | Ketaserin |
| Tacrolimus | **Antidiuretic hormone** |
| **Decongestants** | Vasopressin |
| Ephedrine | **Anti-psychotics** |
| Pseudoephedrine | Haloperidol |
| Phenylephrine | Mesoridazine |
| **Antidepressants/Tricyclic antidepressants** | Chlorpromazine |
| Escitalopram | Ziprasidone |
| Venlafaxine | Anisulpiride |
| Amitriptyline | Thioridazone |
| Citalpram | Doxepine |
| Lithium paroxetine | Ziprasidone |
| Fluoxetine | **Others** |
| Imipramine | Papaverine |
| | Probucol |

*These medicines have been withdrawn worldwide due to the risk of QT prolongation/TdP. The cardiac drugs that tend to prolong QT interval are listed in the table no: 1.4 below[16].

Table No- 1.4: Cardiac Drugs Causing Long QT Interval

| Antiarrhythmic Drugs | Vasodilators/Calcium Channel Blockers |
|---|---|
| Amiodarone | Prenylamine |
| Quinidine | Fenoxedil |
| Sotalol | Bepridil |
| Disopyramide | Terolidine |
| Flecainide | |
| Ibutilide | |
| Dofetilide | |

Below is a list of therapeutic areas of drugs that were listed in the credible meds with known risk, possible risk and conditional risk of TdP which are shown in the figure: 1.5.

Drugs were placed in the respective category after they have shown prolongation of QT interval[22].

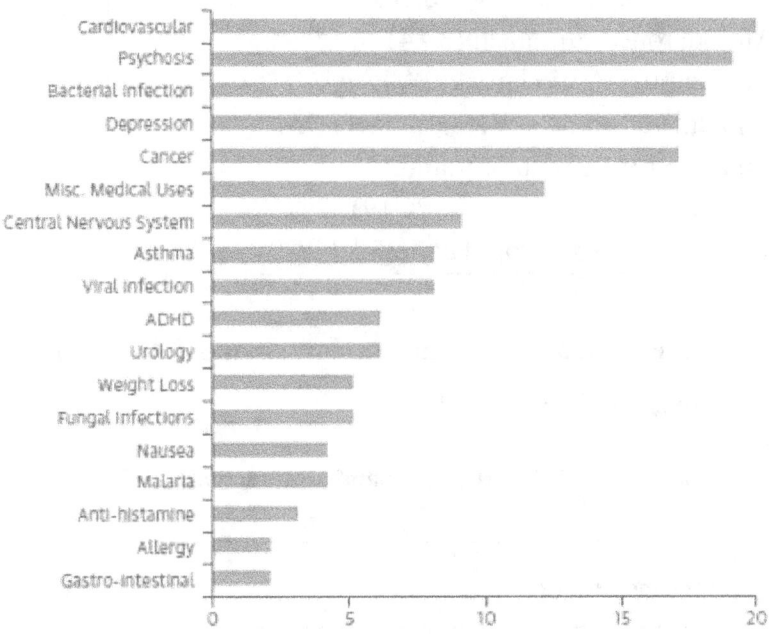

**re no-1.5: Spectrum of Therapeutic Classes Showing the Incidence of QT Prolongation**

## 1.6 Drug Interactions Causing QT Interval Prolongation

Three types of mechanisms through which drug interactions cause long QT interval are[23]:

- Pharmacokinetic drug interactions
- Pharmacodynamic drug interactions
- Electrolyte effects

➢ **Pharmacokinetic interactions:** Some drugs affect the metabolism of other drugs which cause QT prolongation by affecting their metabolism. For example, drugs that inhibit CYP3A4(cytochrome P) enzyme-like macrolides (clarithromycin, azithromycin, erythromycin) and antifungals (fluconazole, ketoconazole, terbinafine) [24], which are listed in the table: 1.5.

Table No- 1.5:  Pharmacokinetic Interactions[44]

| Substrate + inhibitor (CYP450 enzyme) |
|---|
| Amiodarone + metronidazole(3A4) |
| Amiodarone + diltiazem(3A4) |
| Verapamil + Amiodarone(3A4,1A2) |
| Diphenhydramine + Venlafaxine(2D6) |
| Ondansetron + escitalopram(2D6) |
| Citalopram +pantoprazole(2C19) |
| Voriconazole+ metronidazole(3A4,2C9) |

➢ **Pharmacodynamic interactions:** concomitant use of more than one QT-prolonging drug leads to this interaction, listed in table 1.6

Table No- 1.6: Pharmacodynamic Interactions[44]

| Medication combination |
|---|
| Amiodarone+ Risperidone |
| Nicardipine + Ondansetron |
| Diphenhydramine + venlafaxine |
| Fluoxetine +Moxifloxacin |
| Tacrolimus+ Voriconazole |
| Amiodarone +azithromycin |
| Haloperidol+ risperidone |
| Sotalol+ domperidone |

➢ **Effect on Electrolytes:** Diuretics interact with drugs causing QT interval prolongation by causing hypokalaemia. For example, furosemide[24].

## 1.7 Mechanism of Drug-Induced QT Prolongation/TDP:

At the myocardial cellular level, the repolarization period of the myocardial cells takes place principally by efflux of potassium ions. There is diversity in the potassium channel subtypes present in the heart myocytes out of those two currents are most important that take part in repolarization of ventricles, they are IKr(rapid) I.e. delayed rectifier current and IKs(slow) rectifier current[25]. IKr channel is encoded by Human Ether-a-go-go Related Gene (HERG), which is a voltage-gated potassium channel that intervenes the rapid part of the delayed rectifier potassium current[26]. HERG is one of the crucial current accountable for the repolarisation ventricular myocardial cells. HERG and IKr are indistinguishable pharmacologically and are being blocked by methane sulfonanilide and dofetilide which are anti-arrhythmic drugs[27].

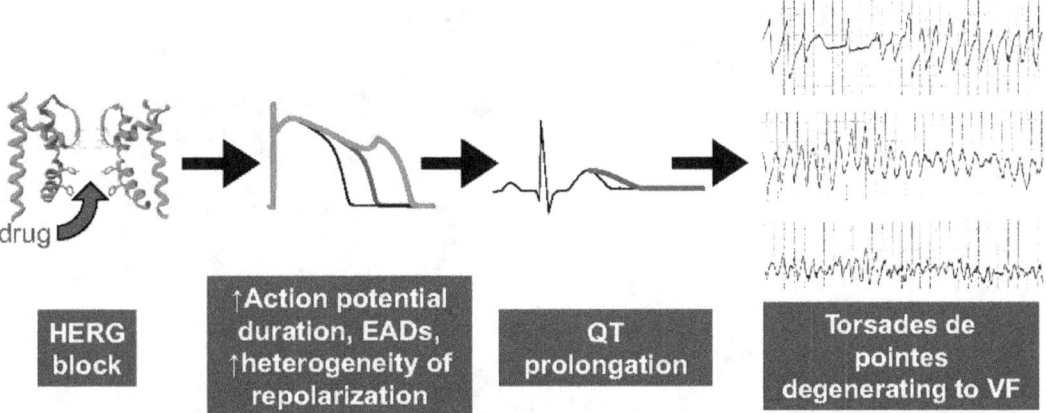

**Figure No- 1.6: Cascade Mechanism of Drug-Induced QT Prolongation.**

Nevertheless, there is an augmented risk for developing new arrhythmias such as TdP, whenever the QTc duration exceeds 500ms due to a drug which causes a rise in 60ms to 70 ms from the previous QTc value[28]. Recently Yang T and co-authors have reported that few drugs that inhibit the IKr potassium current are well known to cause TdP, for example, the drugs such as dovetailed, sotalol, erythromycin were mentioned they are also well known to augment the late sodium current, which may promote to the proarrhythmic effect of the drugs, especially anti-arrhythmic drugs[29]. The potential to obstruct the HERG potassium current, by decreasing IKr rectifier current and causing QT interval to prolong is a universal attribute of QT-prolonging drugs to cause TdP[28].

The delay of repolarisation can lead to sequential activation of an inward depolarization current, also known as an early after depolarisation which may encourage triggered activity of the ventricular myocardial cells, this may induce re-entry of the electrical activity and lead to TdP, such type of activity is induced in His Purkinje fibres and mid-ventricular myocardial cells (M cells)[25]. Compared to IKr channel blockade, M cells show more marked action in the view of prolongation of action potential[25]. Below is the flow chat of arrhythmogenesis of Torsades de pointes and ventricular fibrillation[25] shown in the figure: 1.7.

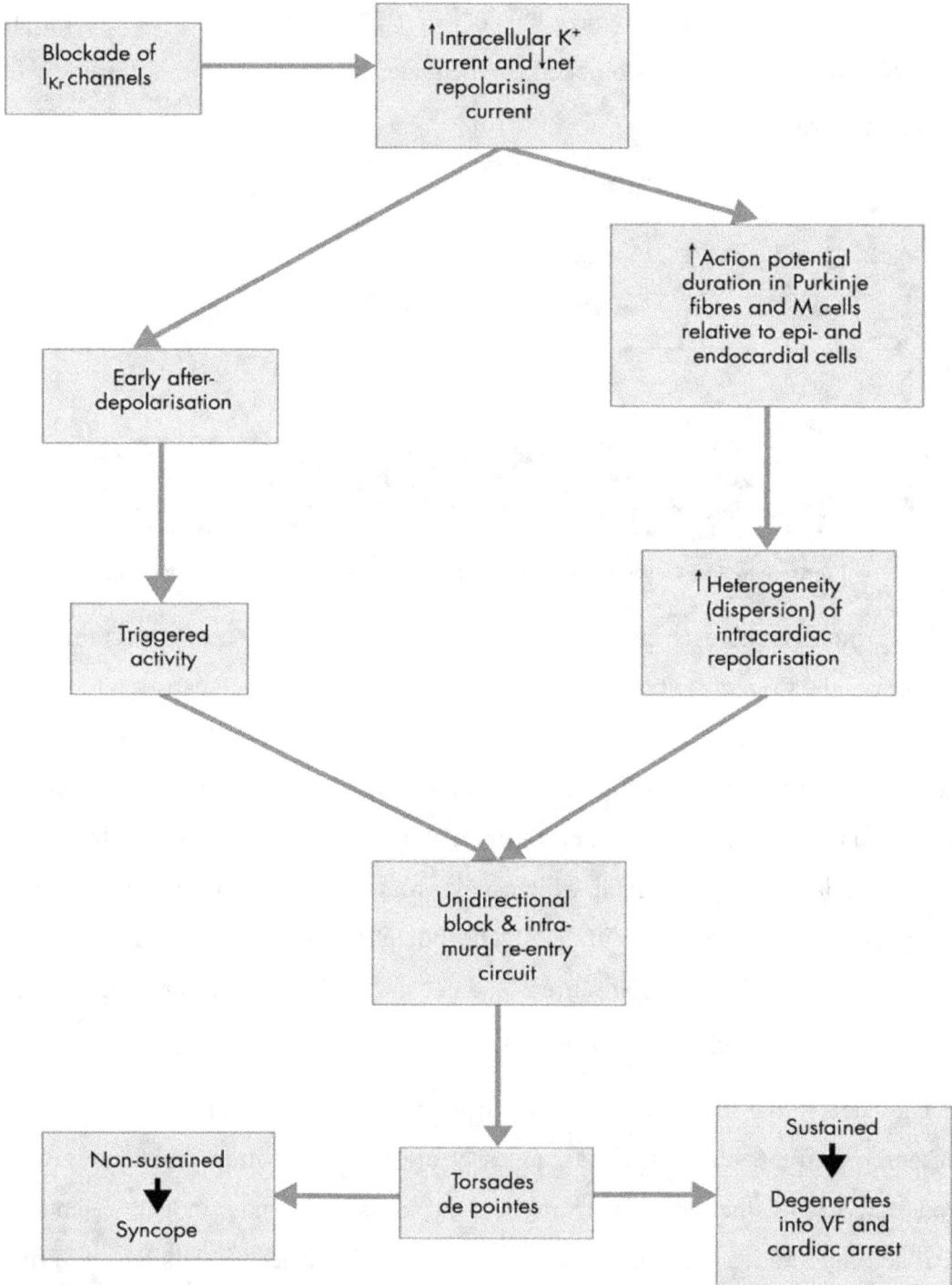

**Figure No- 1.7: Arrhythmogenesis of Torsades De Pointes and Ventricular Fibrillation.**

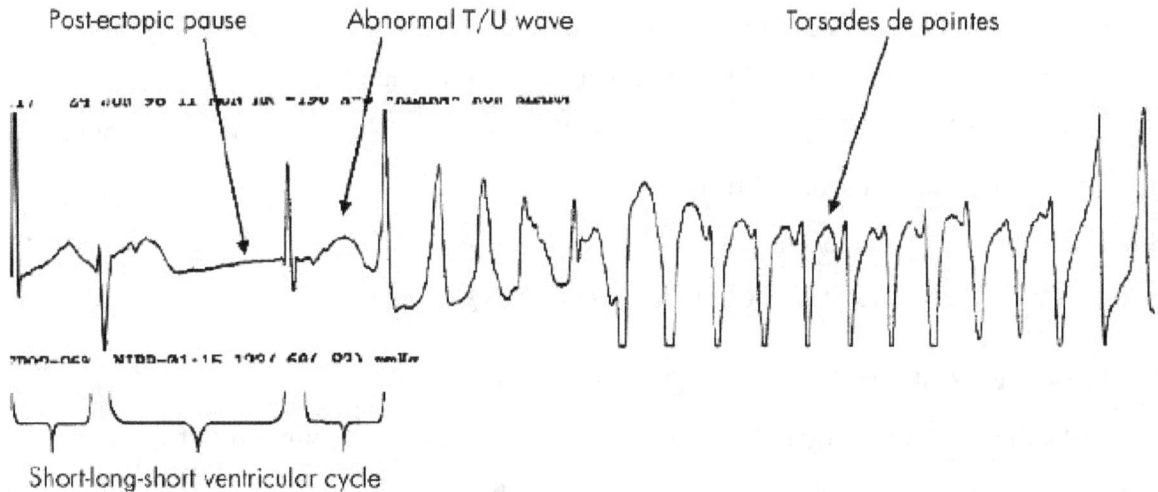

**Figure No- 1.8: Pattern of Torsades De Pointes.**

This is an ECG strip of a patient in whom drug-induced TdP was developed. There is a typical pattern that originates before TdP develops which can be seen in figure: 1.8. Firstly, there are ventricular premature beats I.e. long-short ventricular cycle followed by a pause by sinus beat then pause dependent QT interval prolongation and an elaborated TU wave that truly justifies the term torsion or twisting around a point I.e. baseline of ECG (isoelectric line) in addition to this a careful family and clinical history is required for diagnosis[17].

It has been recommended that after-pause feature of the U wave, in case it is present, maybe a better predictor of TdP induced by drugs than the calculated duration of QT interval[30].

## 1.8 Monitoring an ECG:

According to the American Heart Association practice standards an ECG is recommended for patients who:

- ❖ Have started to take a drug having the potential to prolong the QT interval.
- ❖ Have detected overdose from proarrhythmic drugs.
- ❖ Have first appearance bradyarrhythmia.
- ❖ Have severe hypokalaemia or severe hypomagnesemia.

❖ Patients who have an increase of at least 60ms in their QTc value on ECG after the initiation of QTc prolonging drugs should be monitored.

❖ Patients who have hypothyroidism.

❖ Patients who have previously experienced QT prolongation in past

❖ Any patient who has experienced light-headedness, syncope, palpitations and dizziness should proceed further for investigations[9].

The QTc calculated value should be recorded in not less than 8 to 12 hours after the administration of the potential drugs that cause Long QT interval[31].

## 1.9 Management of QT Prolongation and TDP:

As there are pathological differences in the causes of prolongation of QT interval, so as there are differences in the management[32]. First-line therapy for long QT interval leading to torsades de pointes is instant intravenous administration of magnesium sulphate 2g intravenous bolus over 1-2 mins and repeat dose if the QTc has not decreased[33]. Atropine and isoproterenol have been used successfully in treating the torsades de pointes[34].

Cessation of the drugs causing QT interval prolongation and rectifying the electrolyte abnormalities can be helpful for acquired QT prolongation[35]. In some cases, with bradycardia temporary pacing may be required to shorten the QT interval[35]. In patients who are not retaliating to magnesium sulphate or atropine, ventricular or atrial pacing may be required at the rate of more than 90 beats per minute. Outside hospital settings percutaneous overdrive pacing has found effective in treating TdP[36].

## Long term treatment:

Pharmacological therapy by beta-blockers is usually approached for treating congenital QT syndromes and surgical therapy such as left cervicothoracic sympathectomy is used (37). Cardiac pacing is used to tackle early after depolarizations and bradycardia.[35] Patients who do not tolerate beta-blockers especially due to severe bradyarrhythmia or atrioventricular (AV) block, they must be provided long term pacing.[38] In high-risk individuals, though combination therapy of beta-blockers and pacing, due to noncompliance implantable cardioverter-defibrillator (ICD) backup plan is necessary[39].

By administering serum potassium and spironolactone which in turn causes an increase in potassium levels, is an effective treatment to shorten the prolonged QT interval[40].

Likewise, nicorandil has also successfully shown shortening of QT interval when given in patients with HERG mutations as well as sodium channel blockers have also shown successful treatment[41-43]. In pregnant women, beta-blockers should be taken and should not be withdrawn in pregnancy period and post-partum period[40].

## 1.10 Prevention of QT Interval Prolongation and Torsades De Pointes:

Physicians must involve in an ongoing process of being to be watchful always in evaluating the patient's possibility of developing prolongation of QT and following torsades de pointes, principally in patients who obtain polypharmacy. In summation to that, physicians can review the use of drug utilization review (DUR) systems as the data is a data is accessible for the detection of adverse drug reactions (ADRs) and Drug-Drug interactions causing QT interval prolongation[44].

Assessment of Drug Interactions & Drugs Resulting in QT Interval Prolongation

# Chapter-2
# Review of Literature

**Mika Seto,** *et al,* have done research work on QT-interval prolongation due to medication found in the preoperative evaluation- The study has reported 3 cases in which the prolonged QT interval was improved when the drugs suspected to have caused QT interval prolongation (observed on ECG) were discontinued. One of the cases which were reported the QTc was improved from 511ms to 429 ms when mosapride was discontinued. Risk of TdP is increased when QTc value shows >60ms increase than the previous ECG before administration of the suspected drug[45]. In case 2, the patient's ECG recorded QTc as 473ms indicating QT prolongation and his prescription were containing dasatinib. Soon after discontinuation of dasatinib within 3 days, the QTc had come back to normal i.e. 443ms. Lethal arrhythmia is when QT c is >500ms and mild QT is usually less than <500ms and is not that matter of concern[44].

**Samarendra P,** *et al,* have reported a case on QT prolongation associated with a combination of Azithromycin and Amiodarone. The study has described that the patient was on amiodarone(200mg/d) for atrial fibrillation for a year, she was also taking enalapril and furosemide. Her QTc showed 510ms and QT dispersion 58ms. Treatment for pneumonia was started with azithromycin 500mg on day 1 and 250 mg next 3days. Her ECG showed bradycardia and marked QT prolongation of 660ms and increased QT dispersion of 140ms. This interaction can be concluded by an explanation that Azithromycin has inhibited the CYP3A4 enzyme which is the main metabolising enzyme of Amiodarone resulting in Amiodarone plasma level concentrations thus increase in QT interval prolongation[46].

**Barbara winiowska,** *et al*, have done research work on Drug-drug interactions and QT prolongation as a commonly assessed cardiac effect-comprehensive overview of clinical trials. The study disclosed that the development of torsades de pointes was seen in 2 patients who were treated with a combination of quinidine and amiodarone. Amiodarone caused significant elevation of quinidine i.e 90% elevation of Cmax of quinidine, which led to 41% prolongation of QT interval[47]. They have also described the results of clinical studies with terfenadine either given alone or in combination with CYP3A4 inhibitors such as ketoconazole, itraconazole, clarithromycin, and erythromycin. When terfenadine was given alone QT was not prolonged, but when given in with ketoconazole it has cause highest QT interval prolongation with an increase of >80ms from the previous QTc value, when given with itraconazole>40ms increase in QTc value was seen[48]. Another study investigated that cisapride interaction with QT-prolonging drug of sparfloxacin caused a pharmacodynamic interaction i.e. either synergistic or additive effect, in which both the drugs are QT-prolonging drugs[49].

**Peter J SchwartzMD,** *et al,* have researched Predicting the Unpredictable: Drug-Induced QT Prolongation and Torsades de Pointes The study has analysed that the epidemiology of TdP and death associated with drugs that prolong the QT interval. The incidence ranges greatly for various drugs i.e. (% for quinidine, 1.5% for macrolides and 1% for drugs such as sotalol. Pharmacovigilance companies in German, Sweden, and Italy have found out that approximately there are 0.8 to 1.2 per million-person annual reports of drug-induced QT and TdP[50]. Another study in Berlin found that drug-induced LQT reported in 2.4 per-million person-year for men and 4 per million people in women[50]. A case-controlled study in the Netherlands found out that there is a three-fold increase in the risk of cardiac deaths with QT-prolonging drugs.

**Carlos Rojas Fernandes** *et al,* have researched Current use of domperidone and co-prescribing medications that increase it arrhythmogenic potential among older adults. The study states that those who received a new prescription on domperidone, the estimated dose was 40mg/d and they were co-prescribed with strong or moderate CYP450 inhibitors such as ketoconazole, erythromycin, moxifloxacin, clarithromycin, amiodarone and azithromycin. The authors reported that patients who were receiving a combination of domperidone and ketoconazole resulted in a 3-fold increase in domperidone exposure. This resulted in 15ms QTc value increase from the previous QTc value. The main mechanism underlying could be the inhibition of CYP450 enzyme, which causes the metabolism of domperidone. Thus, resulting in an increased concentration of domperidone. Domperidone is known to inhibit the potassium channels of myocardial cells thus prolonging the ventricular depolarisation and repolarisation[51].

**Sabine MJM, Straus** *et al*, have researched Non-cardiac QTc-prolonging drugs and the risk of sudden cardiac death. In this study, the patients were exposed to non-cardiac QTc prolonging drugs of which few of them are chloroquine, domperidone, clarithromycin, cisapride etc. The exposure of these drugs was a maximum of 7 days. long-term medications such as antipsychotics were exposed to a maximum of 30 days. The relative risk of sudden cardiac death was estimated by the odds ratio using regression analysis[52]. To assess the dose-effect relation trend test was performed. Population attributable risk was calculated percent. The risk was highest in patients taking haloperidol. The patients who were using gastrointestinal drugs risk of cardiac death were highest with domperidone and a three-fold risk in cisapride users. Stratified analysis showed that the risk of cardiac death was high in women than in men. PAR % was about 2% [52].

**Scott R Beach,** *et al*, have researched QT Prolongation, Torsade's de Pointes, and Psychotropic Medications: A 5-Year Update. The study has analysed that the emerging evidence over the past decade has suggested that some SSRIs, particularly citalopram, may have a predictable negative effect on the QTc interval. As recommended by the Food and Drug Administration (FDA) the maximum daily dose of citalopram was to prescribe 40 mg (20 mg in patients with hepatic impairment or those older than 60 years) because of the increased risk of QTc prolongation at higher doses. Among all antidepressants, they have found that citalopram appears to prolong more than other SSRI. Haloperidol also prolongs QT but the risk is higher when given in IV. They have analysed that among antipsychotics ziprasidone and iloperidone are associated with the highest risk of QT and aripiprazole appears safest[53].

**Monica Tarapues,** *et al,* had done research work on Serious QT interval prolongation with ranolazine and amiodarone. The study has revealed in a case that has shown high concentrations of tacrolimus (substrate of CYP3A4) due to inhibition of CYP450 induced by ranolazine[54]. The concurrent use of amiodarone and ranolazine has shown to prolong the QT interval on ECG, this can be due to inhibition of potassium channels together by both the drugs. The study also describes that the concurrent use of amiodarone and metronidazole could result in cardiac toxicity due to CYP3A4 inhibition[54].

**Sang-In Park,** *et al,* have done research work on an analysis of QTC prolongation with atypical antipsychotic medications and selective serotonin reuptake inhibitors using a large ECG record database. In this study, the effects of atypical antipsychotics and SSRIs were compared. It is a 14-drug study that includes 9 atypical antipsychotics i.e. clozapine, amisulpride, aripiprazole, olanzapine, paliperidone, quetiapine, risperidone, sulpiride, and ziprasidone, and 5 SSRIs, which were paroxetine, sertraline, citalopram, escitalopram, and fluoxetine[55]. 63 patients were prescribed either an atypical antipsychotic or SSRI in the QTcB-prolonged group and 768 in the non-prolonged group. QTcB prolongation for the 14 study drugs was significantly different from the positive control drug, cilostazol. There was no significant difference between the frequency of QTcB prolongation for the 14 study drugs and that for the negative control drug, diazepam. Atypical antipsychotic drugs and SSRIs do not significantly prolong the QTc interval. Age showed a significant association with the QTc interval. The risk of QTc prolongation or TdP cannot be ruled out in patients with risk factors including older age, concomitant therapy with other psychotropic drugs, or overdose [55].

**Giuseppe Cocco,** *et al,* had done a research work on Torsades de pointes induced by concomitant use of Ivabradine and Azithromycin: An unexpected dangerous interaction. The study followed a 68 yrs. old case, where the patient had paroxysmal long-lasting palpitations and cardiac syncope and ECG fluctuating between 460ms and 560ms.He was treated with ivabradine 7.5mg, which was well tolerated. After 5 days he was prescribed with azithromycin for acute sinusitis. The syncope occurred without warning and was diagnosed with Torsades de pointes (TdP). As the patient already had long QT administration of Azithromycin resulted in TdP. The CYP450 inhibition is involved in this mechanism[56]. The study also states that ivabradine's risk of inducing TdP is increased when taken with other drugs that block its metabolism[57].

**De Vecchis R,** *et al,* have done research work in 2018 on Acquired drug-induced long QTc: new insights coming from a retrospective study. The study was conducted on patients who are on classes I and III of the Vaughan- Williams classification of antiarrhythmic drugs[58]. Data have been gathered from BCasa Sollievo Della Sofferenza^ Hospital, S.Giovanni Rotondo, and S. Maria del Pozzo Clinic, Somma Vesuviana, Italy) were gathered during the 2008–2017. Patients who were hospitalized or with 450 ms of male or 470 ms of female were taken for research. Among 97 cases only 73 cases were taken into consideration.54 patients had QTc duration > 500 ms. Patients having QTc duration in between 450 and 500 ms did not show any symptoms. 46 patients had symptoms of dizziness, near-syncope, syncope 23 patients did not show any symptoms. QTc duration could not able to predict the symptoms. The main reason for the study is to know the symptoms associated with the duration. Amiodarone-related long QTc is associated with low risk of symptoms compared with other antiarrhythmic drugs. The new-onset long QTc may arise after a few days of exposure to a given drug[59] or within a prolonged period. The risk results from a predisposing genetic substratum are related to cardiotoxic and arrhythmogenic characteristics of a given drug, which do not depend on the extent of QT increase on ECG.

**Bindraban AN,** *et al,* have done the study on Development of a risk model for predicting QTc interval prolongation in patients using QTc-prolonging drugs. The study states that the QTc intervals exceeding 500 ms cause arrhythmias[60]. Patients with a high risk of QT prolongation should avoid QT prolongation drugs[61]. Patients with over 500ms QTc prolongation were considered. ECGs were excluded if patients were younger than 18 years of age at the moment the ECG was recorded, had a QRS complex above 120 ms or if they had a QTc interval of less than 300 ms or more than 600 ms. Among 19,340 ECGs 1343 ECGs were prolonged. The proportion of ECGs with a prolonged QTc interval in patients with a risk score of zero was 2.7%, while in patients with a risk score of 13 or higher the proportion ECGs with prolonged QTc interval was 26.1%. The area under the ROC curve was 0.71 (95% CI 0.68–0.73). The sensitivity and specificity were 0.81 and 0.48. For patients using one or more QTc prolonging drugs, a risk model was developed. 500 ms was chosen as the threshold. The management of drug-induced QTc prolongation includes a balance between the small risk of TdP and sudden cardiac death and the risk of withholding first-line therapies and switching to non-QTc prolonging alternatives. For QTc interval prolongation in patients using one or more QTc prolonging drug, a risk model was developed. By this model reduction of the number of alerts in patients with a low risk of QTc prolongation and improve patient safety by reducing alert.

Rochester MP, *et al,* has researched evaluating the risk of QTc prolongation associated with antidepressant use in older adults: a review of the evidence. The study has described that Antidepressants are widely prescribed medication in older adults which causes QTc prolongation[62]. Patients with >60 were included in the research. Drugs other than antidepressants were not included. Tricyclic antidepressants are used in the treatment of neuropathic pain. TCAs resulted in a 6.9 ms increase in the QTc interval (95% CI 3.1–10.7), with nortriptyline having the greatest increase in QTc at 23.3 ms (95% CI 7.7–38.9 ms), there were only six subjects in the trial. Selective serotonin reuptake inhibitors are the first-line therapy in the management of geriatric depression[63]. The cross-sectional analysis of ECGs of 436 SSRI users demonstrated an over-all QTcf increase of 2.9 ms (90% CI 1.3–4.5). The only agent that caused an increase greater than 10 ms was citalopram, with a mean increase in QTc of 12.8 ms (90% CI 7.3–18.2) in 35 subjects. Serotonin-norepinephrine reuptake inhibitors (SNRIs) are used as second-line agents in the management of geriatric depression. A trial was conducted for 311 patients who are on the duloxetine 60 mg daily. The change in the QTc interval compared with placebo was not significant Venlafaxine affected a mean increase in  QTc interval of 10.6 ms (95% CI 9.2–12.0)[63]. SSRIs increased the risk for sudden death tricyclic antidepressants increased risk but failed to reach statistical significance. Monitoring of electrolytes is important for monitoring patients on Qtc prolongated drugs. TCAs and citalopram cause the risk for QT prolongation in older adults whereas the other SSRIs and SNRIs do not cause any significant risk.

Galia Jackobson, *et al,* has done the research work on reckless administration of QT interval-prolonging agents in elderly patients with drug-induced torsade de pointes. The study discloses that QT prolongation leads to Tdp[64]. Elderly patients use diuretics for congestive heart failure which causes hypomagnesemia, hypokalemia[65]. 67 patients case reports were used for the study. 5 patients case study was excluded as it was not independently responsible for Qt prolongation[66]. Demographic details of the patient were collected. Among 61 reports on drug-induced Tdp of elderly patients, the mean range was 84.3 ± 3.9 years (median 83 years, range 80–97 years). In 39 (63.9%)patients baseline, QTc interval and the mean QTc interval was 444.2 ± 29.8 ms (median 446 ms, range 385–510 ms). antiarrhythmic agents (n = 19,31.1%) and antibiotics (n = 18, 29.5%)followed by psychotropic drugs (n = 16,26.2%) and other agents were the most prevalent drugs that triggered TdP.The most reckless administration appeared when antiarrhythmics and antibiotics were given (25%). Followed by (18.8%) antiarrhythmics and psychotropic drugs, one (6.3%) patient was treated with antibiotics and psychotropic drugs and eight (50.0%) patients were treated with other combinations of QT interval prolongation agents. In elderly 50% of Qt prolongation was found due to reckless administration of Qt prolongation drugs. Physicians should particularly avoid prescribing two or more QT interval-prolonging agents simultaneously in the elderly population.

**Tisdale JE,** *et al,* have done the research work on Prevalence of QT Interval Prolongation in Patients Admitted to Cardiac Care Units and Frequency of Subsequent Administration of QT Interval-Prolonging Drugs. The study reveals that the TdP caused by cardiac arrest is an uncommon but catastrophic event in hospitalized patients[28]. Patients admitted to cardiac care units may be at especially high risk of drug-induced QT interval prolongation and TdP due to the presence of underlying heart disease, electrolyte abnormalities. Demographic details, past medical history, daily progress notes, medication administration records, laboratory data, ECGs, telemetry monitoring strips, and diagnostic reports were collected. The age group of >18 years and 'outpatient were excluded. QTc interval >500 ms was considered abnormally high in both males and females[45]. Among 1159 patients 900 patients were included in the study due to exclusive criteria. 790 had normal sinus rhythm,76 had atrial fibrillation and 34 had frequent ventricular premature depolarizations or premature atrial contractions. Common QT interval-prolonging drugs given to patients were intravenous or oral amiodarone, a fluoroquinolone or macrolide antibacterials, methadone, ranolazine and ziprasidone. Many patients received intravenous amiodarone before admission.25% of patients had QTc interval prolongation on admission, 20% of patients had QTc interval >500 ms, more than one-third of the patients received a QT interval-prolonging drug during hospitalization, and over 40% of patients with QTc interval >500 ms, later on, received a QT interval-prolonging drug. QTc interval prolongation occurred in one-third of patients who presented with QTc interval prolongation and who received QT interval-prolonging medications, and in >50% of patients who presented with QTc interval >500 ms who were given Qt prolonging drugs. According to this study, Qtc interval prolongation was common in patients admitted to cardiac units.

Allen LaPointe NM, *et al,* has performed the study on Frequency of high-risk use of QT-prolonging medications. The study states that the research was done on 2 million health plan members from 10 health maintenance organizations. Age >65 years, female, history of coronary artery disease, or myocardial infarction, history of heart failure, or left ventricular dysfunction, and conduction system disease or heart block without evidence of a pacemaker were included in the study. Among 228 550 patients (11.4%) had 691 263 claims for QT-prolonging medications. Of these, 182 047 patients (79.7%)had 304 272 claims for a higher-risk QT-prolonging drug. Potential drug interactions involving QT-prolonging drugs occurred 48465 times in 10 415 patients (4.6% of patients with a claim for a QT-prolonging medication).3653 were females. Amon 48 465 potential drug interactions, 42 961(88.6%) occurred in patients with one or more of the risk factors.39% of cases are of more than 1 QT-prolonging drug, and 38% are due to combination of a QT-prolonging drug with the drug that inhibits its metabolism[67].In this study, Amitriptyline caused more drug interactions of QT-prolongation QT-prolonging medications occurred in 4.6% of patients who were receiving a QT-prolonging medication.

Makaryus AN, *et al,* have done the study on Effect of ciprofloxacin and levofloxacin on the QT interval: is this a significant "clinical" event? The study was done to research whether the Ciprofloxacin and levofloxacin may cause QT prolongation. In this study, 27 patients received levofloxacin and 11 ciprofloxacin with the standard doses[68]. QT prolongation was calculated using bazett's formula[69]. Patients with atrial fibrillation, any type of arrhythmias were excluded. There was a slight QT prolongation among patients who received levofloxacin but there was no large increase in QT interval. There was no prolongation in patients who received ciprofloxacin. There was no significant change in the mean value of the QT interval. Neither levofloxacin nor ciprofloxacin significantly prolonged the mean QTc interval over baseline. When electrolyte deficiencies in one of the patients were taken into consideration, this is true for the longest QTc interval. These 2 drugs didn't show any appreciable effect on QT prolongation in this study. This study is not a very "clinically" significant one with the use of ciprofloxacin and levofloxacin when an accurate objective assessment of risk is undertaken.

Kim A, *et al,* has done research work on A thorough QT study to evaluate the QTc prolongation potential of two neuropsychiatric drugs, quetiapine and escitalopram, in healthy volunteers. The study reveals that some antipsychotics and antidepressants are associated with QT prolongation[70]. To healthy 40 volunteers placebo, 400 mg moxifloxacin as a positive control, 20 mg escitalopram,100 mg quetiapine were given. The moxifloxacin washout period was 1 week and for quetiapine, escitalopram was 2 weeks. Blood samples were collected after administration of doses to know the plasma concentrations. ECG was recorded before and after administration of doses. Among 40 individuals 33 completed the study. Moxifloxacin was rapidly absorbed, Tmax was 2 h after dose. The maximum mean $\Delta\Delta$QTcI, after moxifloxacin was 16.2 ms, occurred at 3 h after dose. Escitalopram was slowly absorbed and the $\Delta\Delta$QTcI exceeded 10ms for quetiapine and escitalopram. QT interval was prolonged by quetiapine and escitalopram. QT prolongation by quetiapine was observed around Tmax in this study and at higher doses, there will be more QT prolongation. citalopram and escitalopram showed a dose-dependent increase in QTcF and changes in QTcF after citalopram administration is more than that of escitalopram administration. QT prolongation occurs with the increase in the doses.

Curtis LH, *et al,* has performed the study on Prescription of QT-prolonging drugs in a cohort of about 5 million outpatients. The study states that many cardiac and noncardiac drugs can cause QT prolongation[71]. Concurrent use of 2 or more QT-prolonging drugs can cause QT prolongation retrospective cohort study was conducted in 4,825,345 subjects aged 18 years or older.>18 age group were excluded from the study. Patients with the same family unit were excluded. 50 drugs were identified that causes QT-prolongation, and 26 drugs inhibit the metabolic clearance of QT-prolonging drugs. Half(47.4%) of all patients were prescribed with clarithromycin or erythromycin which causes QT prolongation.40% were prescribed with antidepressants. 7249 patients(0.7%) are prescribed with the overlapping combinations of three or more drugs. 76,118 (74%) were women. Most of the drugs that causing QT-prolongation are antibiotics and antidepressants.

Bush SE, *et al,* has performed the study on Effects of concomitant amiodarone and haloperidol on Q-Tc interval prolongation. The study has described that the amiodarone is to treat both atrial and ventricular tachyarrhythmias. It causes low torsades de pointes than other antiarrhythmic agents[72], but given I'm combination with other QT-prolonging drugs will increase QT prolongation[73]. Haloperidol is used to treat symptoms of psychosis and delirium[74]. Adult patients who were admitted in a tertiary referral teaching hospital were included for this retrospective descriptive analysis that are given both haloperidol and amiodarone.57 patients were given with both medications and amiodarone was given first in those patients. Only 49 patients were included in the study as for the remaining 9 patients there was no ECG. 49 patients were given 381 distinct amiodarone–haloperidol exposures. 16 patients (28%) were given one additional Q-T interval-prolonging drug, two additional Q-T interval prolonging drugs during 13 exposures (3%) (3 patients, 5%), and three or more additional Q-T interval-prolonging drugs during 11 exposures (3%) (2 patients, 4%). There was a mean increase of QTc interval after exposure of 9.8 msec after the administration of haloperidol. There was an increase in QTc prolongation when these both drugs are given.

# Chapter-3
# Need for the Study

1. A study report relates that QT prolongation leads to sudden cardiac death where it is majorly suffering in countries like United States, United Kingdom, Australia, India, Europe[75].

2. Indian research says that the prevalence of prolongation was found to be up to 34.1% in a selected number of patients in various medical conditions like kidney disease, chronic liver disease, haemorrhage cerebrovascular accident and heart failure[76].

3. As per the literature survey, it was understood that the QT interval prolongation is predominant in Indian emergency medical patients[76].

4. In light of this perspective, there is a need to focus on assessment on QT interval prolongation due to specific drugs and drug interactions in a specific regional area patient.

Assessment of Drug Interactions & Drugs Resulting in QT Interval Prolongation

# Chapter-4
# Aim and Objectives

## Aims:

The main aim of our study is

- ➤ To identify potential drug-drug interactions and drugs causing QT interval prolongation cardiac outpatients
- ➤ To find the prevalence of potential drug-drug interactions and drugs causing QT interval prolongation in The Department of Cardiology,
- ➤ To make a list of most common potential drug-drug interactions as well as drugs causing long QT interval in the cardiac outpatient department and
- ➤ To determine the risk factors associated with it.

## Objectives:

### Primary objective:

To assess the proportion of patients with QT interval prolongation among all the patients who are screened in cardiac outpatients over 6 months.

### Secondary objective:

- ➤ To identify the potential drug-drug interactions that prolong QT interval
- ➤ To identify drugs that prolong QT.

Assessment of Drug Interactions & Drugs Resulting in QT Interval Prolongation

# Chapter-5

# Plan of Work

The study was performed in the following steps:

- ➢ Literature survey
- ➢ Collection of the consecutive cases in the cardiac outpatient setting.
- ➢ Analysis of prescription of each case for serious drug interactions and drugs causing QT interval prolongation their associated risk factors, co-morbidities.
- ➢ Collection of Socio-demographic details such as age, gender, height, weight, etc. medical history, clinical & treatment details.
- ➢ Collection of data needed for the study from ECG and 2D ECHO reports, and lab reports such as serum electrolytes (hypokalaemia, hypomagnesemia, hypocalcaemia).
- ➢ Analysis of prescriptions based on details like the number of drugs, names of individual drugs, dose, dosage form, dosing schedule &duration.
- ➢ Analysis of adverse drug reactions using Naranjo's causality assessment scale and Drug-drug interactions using Micromedex solutions.

Assessment of Drug Interactions & Drugs Resulting in QT Interval Prolongation

# Chapter-6
# Methodology

---

**Study Site**:

The study was carried out at the Care Hospitals, Banjara Hills, Road no:1, Hyderabad, a Tertiary Care Centre.

**Study Design**:

This is a non-randomized observational prospective cohort study. The data will be collected prospectively from the cardiac outpatient department.

**Study Duration:**

6 Months.

**Ethical Considerations**:

The study was approved by the institutional ethics committee of Care Hospital, Banjara Hills, Hyderabad (Protocol no: ECR/49/Inst./AP/2013/RR-16/EC approval no-A 14). All consents to be taken in local languages and patient/family to be fully explained about the nature of the study.

**Study Criteria:**

Table No- 6.1: Inclusion and Exclusion Criteria

| Inclusion Criteria: | Exclusion Criteria: |
|---|---|
| Age groups above 18 years are included. All patients in the cardiac outpatient. | The age group below 18 years of age Patients who have congenital long QT syndrome. Patients who have electrolyte imbalances. |

The present study involves the detection of QT prolonged cases which may be caused due to drugs, drug interactions, electrolyte imbalance, which was prescribed in Care Hospitals, Banjara Hills.

The study involves the following steps.

1.  Collection of prescriptions.
2.  Recording the prescription details in prescribed format proforma (Proforma enclosed in appendix).
3.  Analysing the prescription and its lab investigations to identify QT prolonged cases.

## Collection of Prescriptions:

A total of 1123 prescriptions were collected in the electrophysiology department, care hospital, Banjara hills for 6 months.

## Recording the Prescription Details in Prescribed Format Proforma (proforma enclosed in appendix):

Patient details of each case sheet-like prescription and lab investigations were noted in the proforma. Proforma includes age, name, sex, diagnosis, chief complaints, drugs, Drug-drug interactions, lab investigations.

## Analysing the Prescription and its Lab Investigations to Identify QT Prolonged Cases:

The patient prescriptions and lab data (such as ECG, thyroid profile, electrolytes) were analysed for QT prolongation and its causative factors have been identified and noted in the proforma.

## Statistical Analysis:

Continuous variables will be represented as mean and standard deviation where data follows a normal distribution, otherwise as median with range. Categorical variables will be represented as frequencies and percentages. Data will be analysed using the R ratio.

# Chapter-7

# Results & Discussion

This is a non-randomised observational prospective study in which we were able to evaluate the number of drugs induced QT interval prolongation and drug interactions causing long QT cases. A total number of cases (N=1122) were screened, during the study period. The data of the patients were categorised according to age, gender, other causes of QT prolongation, diagnosis, list of drugs, drug interactions, therapeutic class of drugs, mechanisms of interactions and causality assessment. The percentages were calculated according to the collected data.

## 7.1 Proportion of QT Interval Cases:

Among the 1122 cases screened, normal QT interval cases were found to be n=1075, whereas the prolonged QT intervals cases were found to be n=47 which is summarised in table 7.1 and figure 7.1.

Table No- 7.1: Proportion of Prolonged QT Interval Cases among Overall Cohort.

| S. No. | Study Population | Number (n) | Percentage (%) |
|--------|------------------|------------|----------------|
| 1. | Normal QT interval cases | 1075 | 96% |
| 2. | Prolonged QT interval cases | 47 | 4% |
| 3. | Total no of the collected cases | 1122 | 100% |

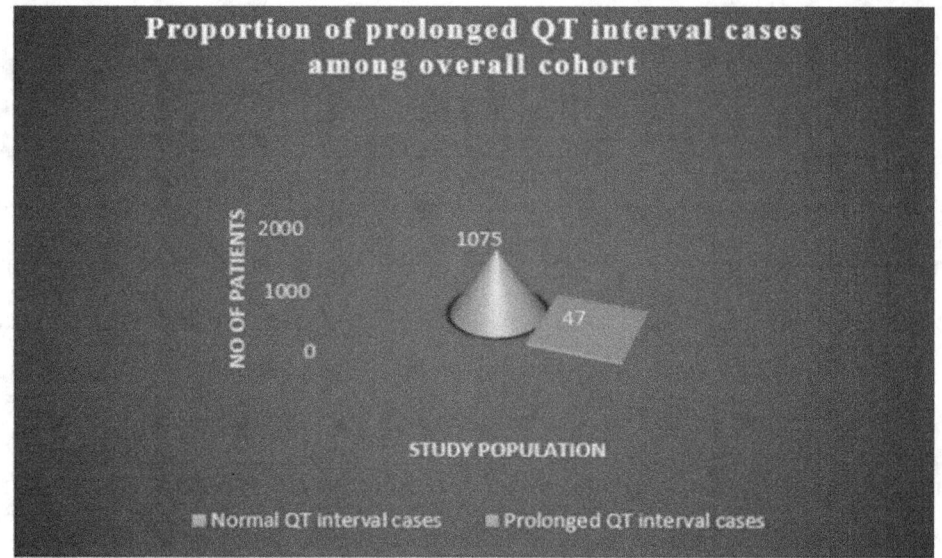

**Figure No-7.1: Proportion of Prolonged QT Interval Cases**

The percentage of the prolonged QT interval cases was found to be 4% and the percentage of normal QT interval cases was 96% which is summarised in table 7.1 and figure 7.2

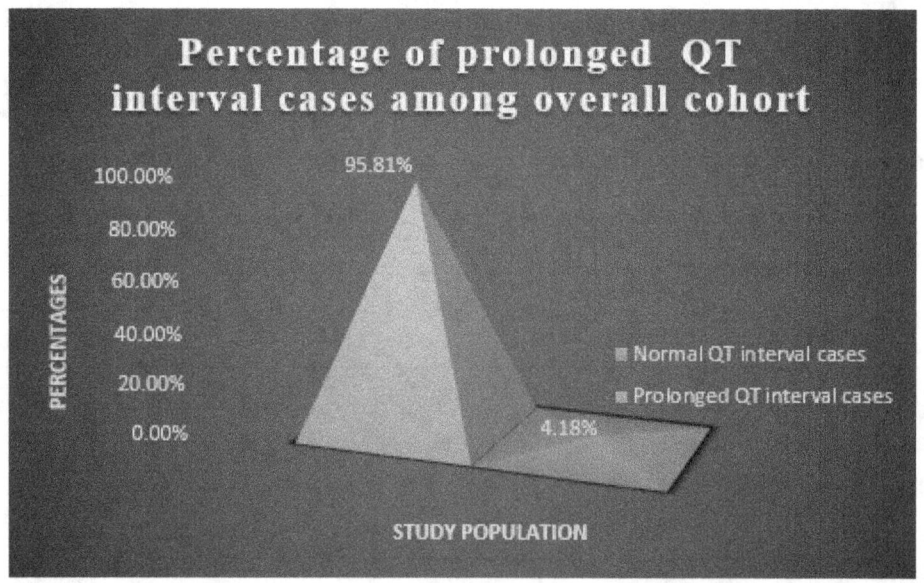

**Figure No- 7.2: Percentage of Prolonged QT Interval Cases**

## 7.2 Gender Distribution:

In the performed study, the 47 cases of QT prolongation were due to various causes and it was observed that the male population (n=31) was higher than the female population (n=16) which is summarised in table 7.2 and figure 7.3.

**Table No- 7.2: Gender Distribution among Various Causes of QT Prolongation**

| S. No. | Gender | Number (n) | Percentage (%) |
|--------|--------|------------|----------------|
| 1. | Male | 31 | 66% |
| 2. | Female | 16 | 34% |
| 3. | Total | 47 | 100% |

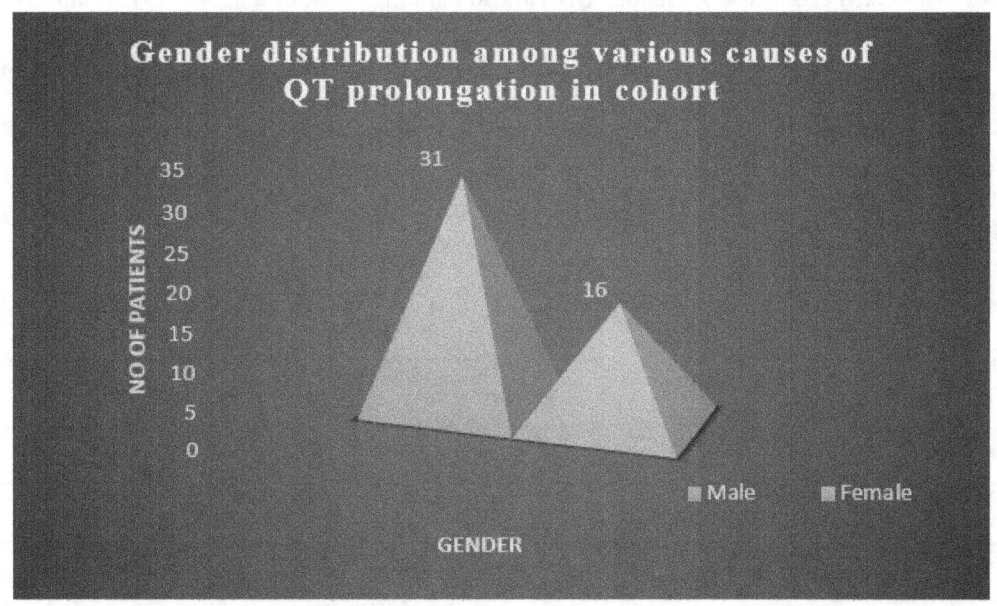

**Figure No- 7.3: Gender Distribution among Various Causes of QT Prolongation**

The percentage of the male population was 66% and the female population was 35% which is summarised in table 7.2 and figure 7.4.

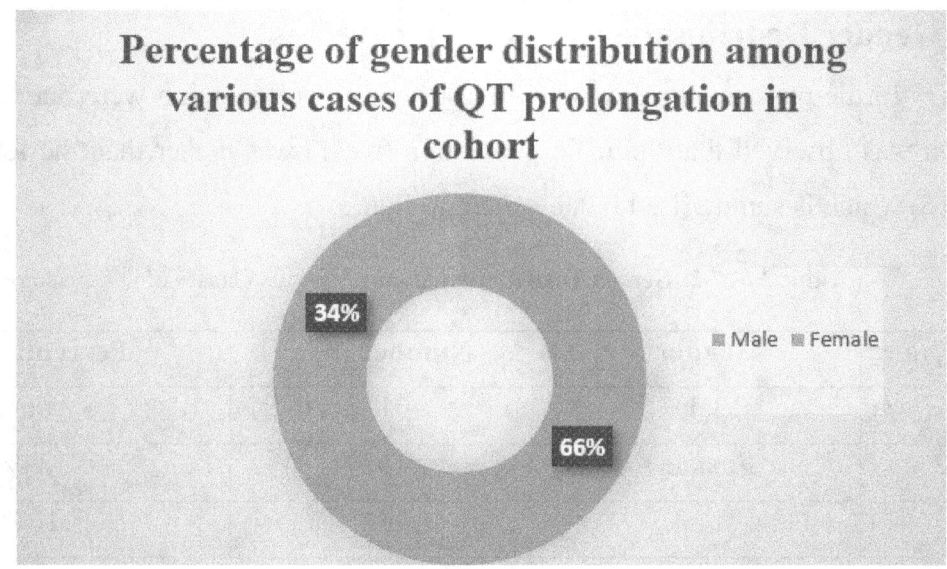

**Figure No- 7.4: Percentage of Gender Distribution among Various Cases of QT Prolongation**

From the n=47 QT prolonged cases, n=27 cases were observed to have prolonged the QT interval due to drugs and drug-drug interactions. Out of these 27 cases, the male population have found to be n=18 and female population have found to be n=9, with male predominance which is summarised in table 7.3 and figure 7.5.

**Table No- 7.3: Gender Distribution among Drug-Induced QT Prolongation**

| S. No. | Gender | Number(n) | Percentage (%) |
|--------|--------|-----------|----------------|
| 1. | Male | 18 | 67% |
| 2. | Female | 9 | 33% |
| 3. | Total | 27 | 100% |

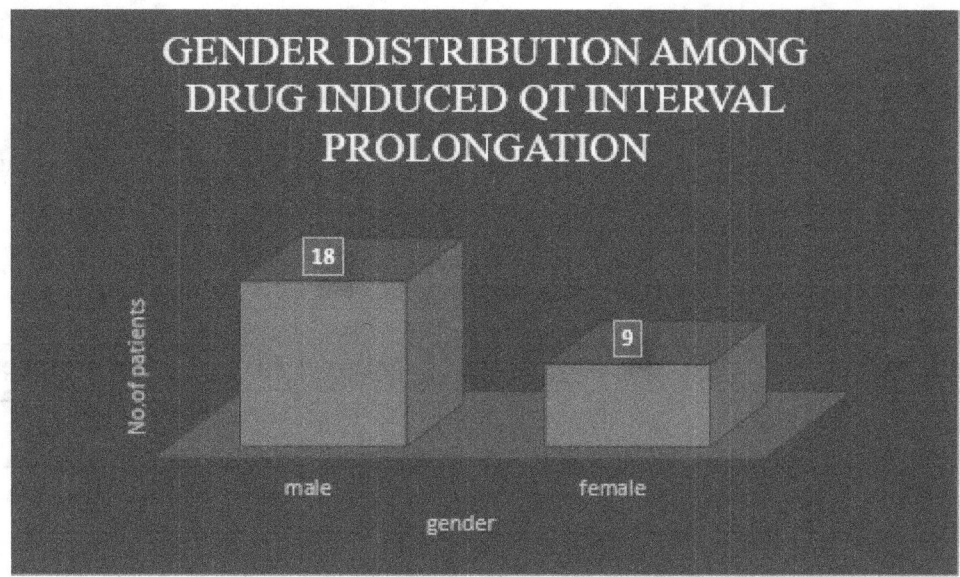

**Figure No- 7.5: Gender Distribution among Drug-Induced QT Prolongation**

The percentage of the male population was observed to be 67% and the female population was observed to be 33% with male dominance which is summarised in table 7.3 and figure 7.6.

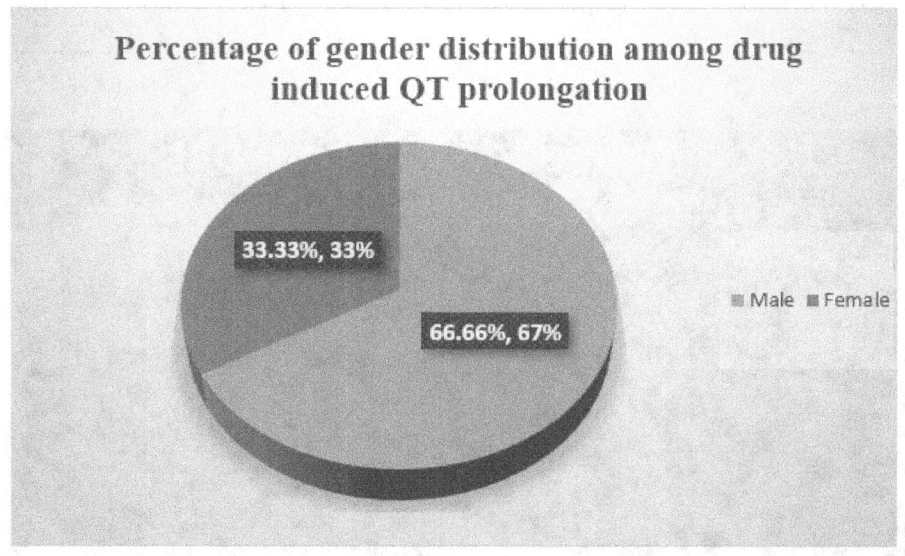

**Figure No- 7.6: Percentage of Gender Distribution among Drug-Induced QT Prolongation**

## 7.3 Age Distribution:

In the conducted study, according to the age group, the highest number of patients who have observed drug-induced QT prolongation were of 51-60 (n=9) and least number was found within the age group of 41-50 (n=1) which is summarised in table 7.4 and figure 7.7. The mean age of patients included in the study is $57 \pm 15$.

The graph shows the highest prevalence in the age group 51-60 and the least prevalence in the age group 41-50.

Table No- 7.4: Age Distribution among Drug-Induced QT Prolongation Cases

| S. No. | Age distribution | Number (n) | Percentage % |
|--------|------------------|------------|--------------|
| 1. | 21-30 | 3 | 11.11% |
| 2. | 31-40 | 3 | 11.11% |
| 3. | 41-50 | 1 | 3.7% |
| 4. | 51-60 | 9 | 33.33% |
| 5. | 61-70 | 5 | 18.51% |
| 6 | 71-80 | 3 | 11.11% |
| 7. | 81-90 | 3 | 11.11% |
| | Total | 27 | 100% |

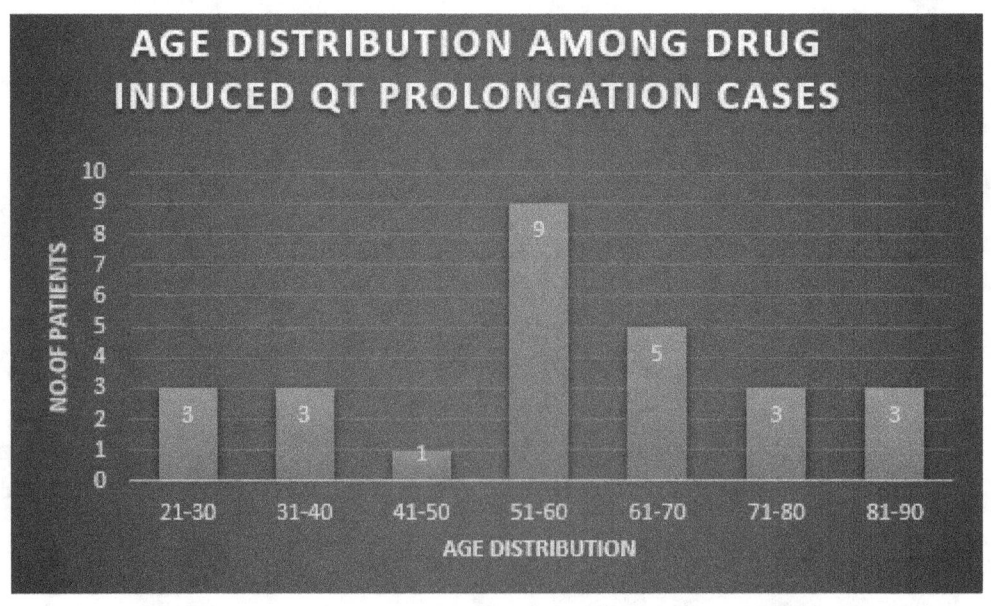

Figure No- 7.7: Age Distribution in Drug-Induced QT Prolongation

The percentage of highest prevalent age group was observed to be 33.33% and the least prevalent age group was observed to be 3.7% among the QT prolongation population which is summarised in table 7.4 and figure 7.8.

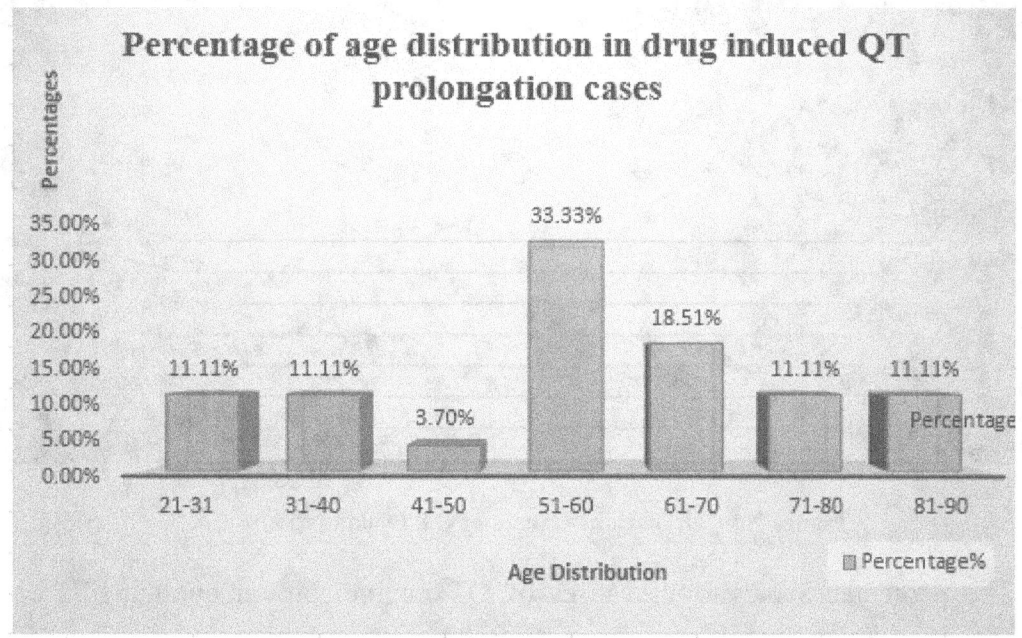

Figure No- 7.8: Percentage of Age Distribution in Drug-Induced QT Prolongation Cases

## 7.4 Various Causes of QT Prolongation:

In the present study 5 major causes that lead to QT interval prolongation were reported which includes drugs, drug-drug interactions, QRS complex widening, electrolyte imbalance and bradycardia which has been summarised in table 7.5 and figure 7.9.

Table No- 7.5: Various Causes of QT Prolongation

| S. No. | Causes | Number (n) | Percentage % |
|---|---|---|---|
| 1. | Drug-induced QT prolongation | 12 | 25.53% |
| 2. | Drug interaction induced QT prolongation | 15 | 31.91% |
| 3. | QRS widening causing QT prolongation | 16 | 34.04% |
| 4. | Bradycardia causing QT prolongation | 2 | 4.25% |
| 5. | Electrolyte imbalance causing QT prolongation | 2 | 4.25% |
| | Total Cases | 47 | 100% |

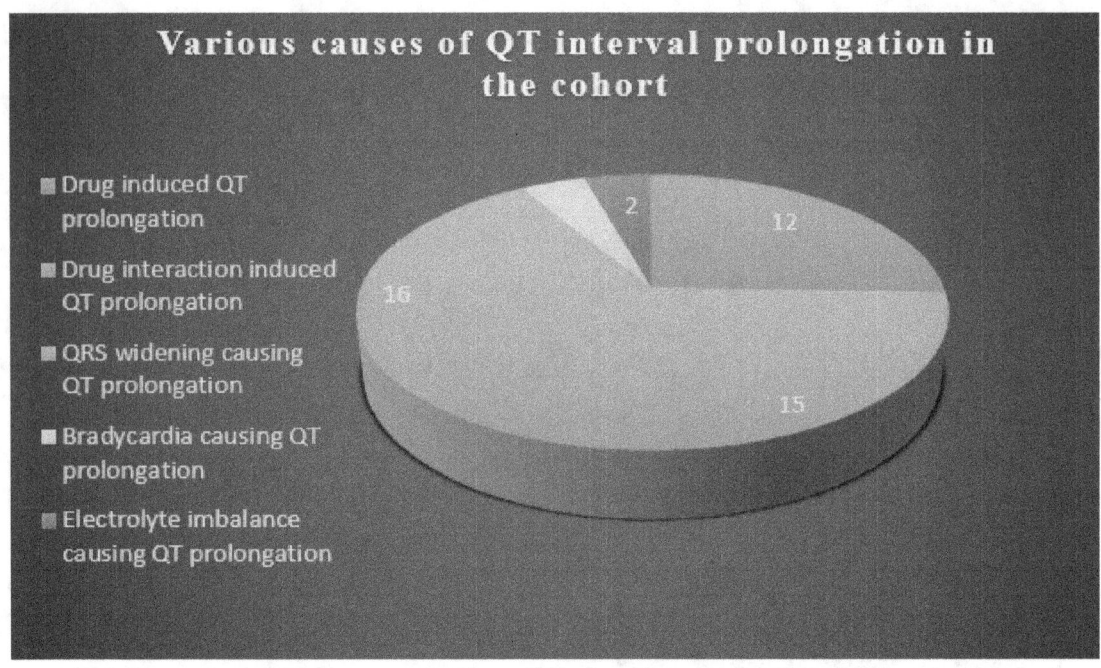

**Figure No- 7.9: Various Causes of QT Prolongation**

The percentages of various causes of QT interval prolongation in the cohort are summarised in table 7.5 and figure 7.10.

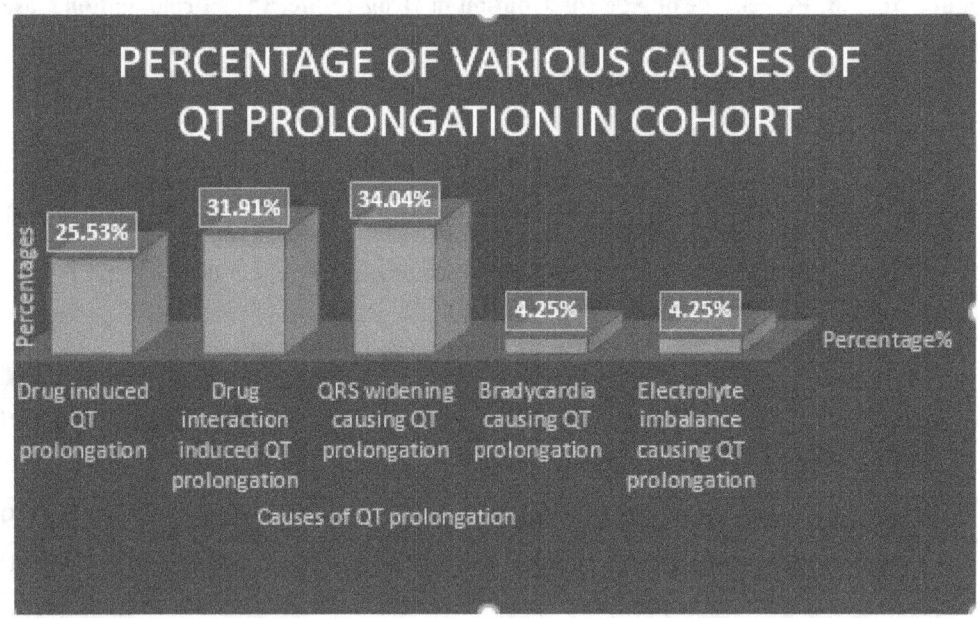

**Figure No- 7.10: Percentage of Various Causes of QT Prolongation**

**Table No-7.6: Co-morbidities in Patients with Drugs Causing QT Interval Prolongation**

| S. No | Causes | Diagnosis | Number of Repeated Diagnoses | Percentage % |
|---|---|---|---|---|
| 1. | Drugs & | Diabetes Mellitus | 11 | 17.10% |
| 2. | Drug | Hypertension | 14 | 21.80% |
| 3. | Interactions | Hypertrophic Cardiomyopathy | 3 | 4.60% |
| 4. | | Ventricular Tachycardia | 3 | 4.60% |
| 5. | | LV Dysfunction | 8 | 12.50% |
| 6. | | Coronary Artery Disease | 7 | 10.90% |
| 7. | | Obstructive Sleep Apnea | 5 | 7.80% |
| 8. | | Dilated Cardiomyopathy | 5 | 7.80% |
| 9. | | Myocardial Infarction | 6 | 9.30% |
| 10. | | Atrial Flutter | 1 | 1.50% |
| 11. | | Atrial Fibrillation | 1 | 1.50% |
| | | Total | 64 | 100% |

## 7.5 Co-Morbidities:

The diagnosis found from these patients may be with individual diagnosis or with co-morbidities.

The listed conditions are isolated with a specific diagnosis, so here in the given table, the number of diagnostic conditions will appear more than the number of patients, where there is no confusion to understanding the list of diagnosis when it is compared with the number of patients, which we have summarized in table 7.6 and figure 7.11.

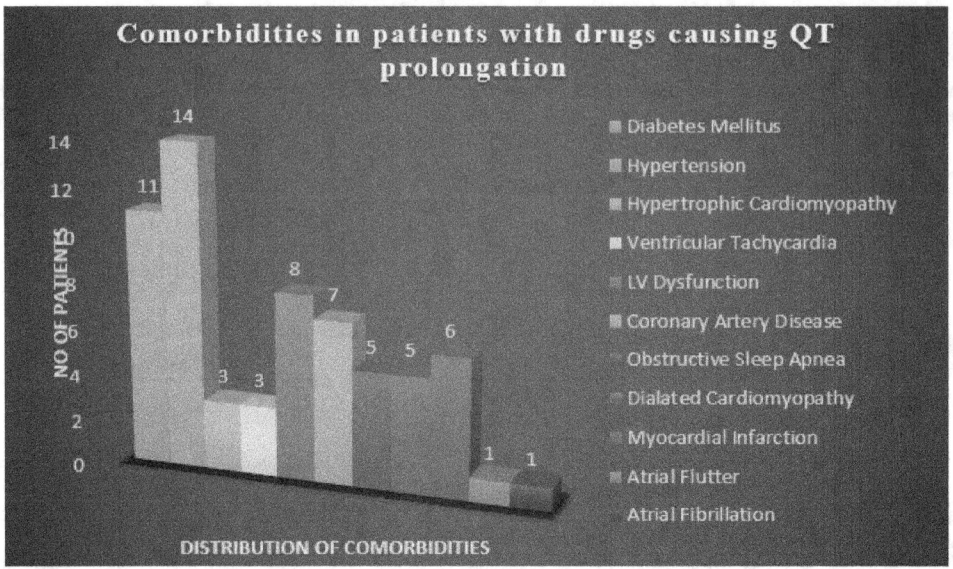

**Figure No- 7.11: Co-morbidities in Patients with Drugs Causing QT Interval Prolongation**

The percentage of the diseases present in patients having Drug-induced and drug interaction induced QT prolongation is shown in the figure: 7.12.

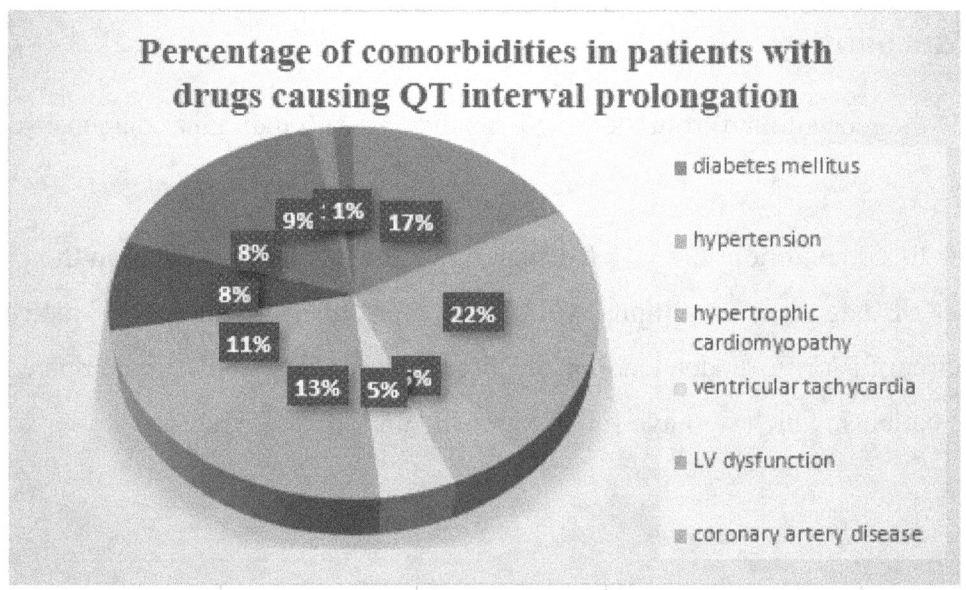

**Figure No- 7.12: Percentage of Co-morbidities in Patients with Drugs Causing QT Interval Prolongation**

Other than drugs, there are other causes in which we have enlisted the comorbidities summarized in the table: 7.7 and figure: 7.13

Table No- 7.7: Co-morbidities in Patients with Other Causes of QT Prolongation

| S. No. | Cause | Diagnosis | N | Percentage% |
|--------|-------|-----------|---|-------------|
| 1. | QRS widening | Hypertension | 10 | 30.30% |
| 2. | Electrolyte imbalance | Diabetes Mellitus | 4 | 12.12% |
| 3. | Bradycardia | Coronary Artery Disease | 3 | 9.09% |
| 4. | | Hypertrophic Cardiomyopathy | 2 | 6.06% |
| 5. | | Ventricular Tachycardia | 3 | 9.09% |
| 6. | | Atrial Fibrillation | 5 | 15.15% |
| 7. | | Myocardial Infarction | 1 | 3.03% |
| 8. | | LV Dysfunction | 3 | 9.09% |
| 9. | | Dilated Cardiomyopathy | 2 | 6.06% |
| Total | | | 33 | 100% |

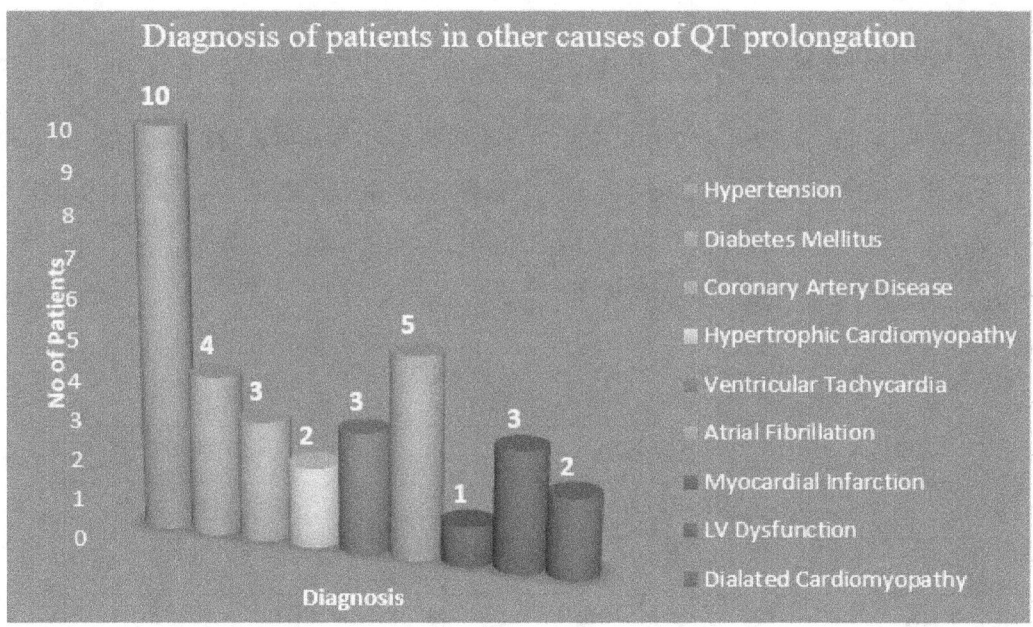

Figure No- 7.13: Comorbidities in Patients with Other Causes of QT Prolongation

The percentages of these comorbidities have been calculated in the figure no: 7.14

**Figure No- 7.14: Percentage of Co-morbidities in Patients with Other Causes of QT Prolongation**

## 7.6 Drugs Causing QT Prolongation:

The study gives information on the drugs that have caused QT interval prolongation which was enlisted in table 7.8 and the most common drugs that have caused QT interval prolongation in figure 7.15. In our study, the most frequent adverse drug reaction (ADR) of the QT prolongation was due to ivabradine.

**Table No- 7.8: Drugs Causing QT Prolongation**

| S. No. | Drugs | N | Percentage% |
|--------|-------|---|-------------|
| 1. | Escitalopram | 2 | 16.66% |
| 2. | Duloxetine | 1 | 8.33 |
| 3. | Terbutaline | 1 | 8.33 |
| 4. | Disopyramide | 2 | 16.66% |
| 5. | Ivabradine | 4 | 33.33% |
| 6. | Ranolazine | 1 | 8.33 |
| 7. | Clarithromycin | 1 | 8.33 |
| | Total | 12 | 100% |

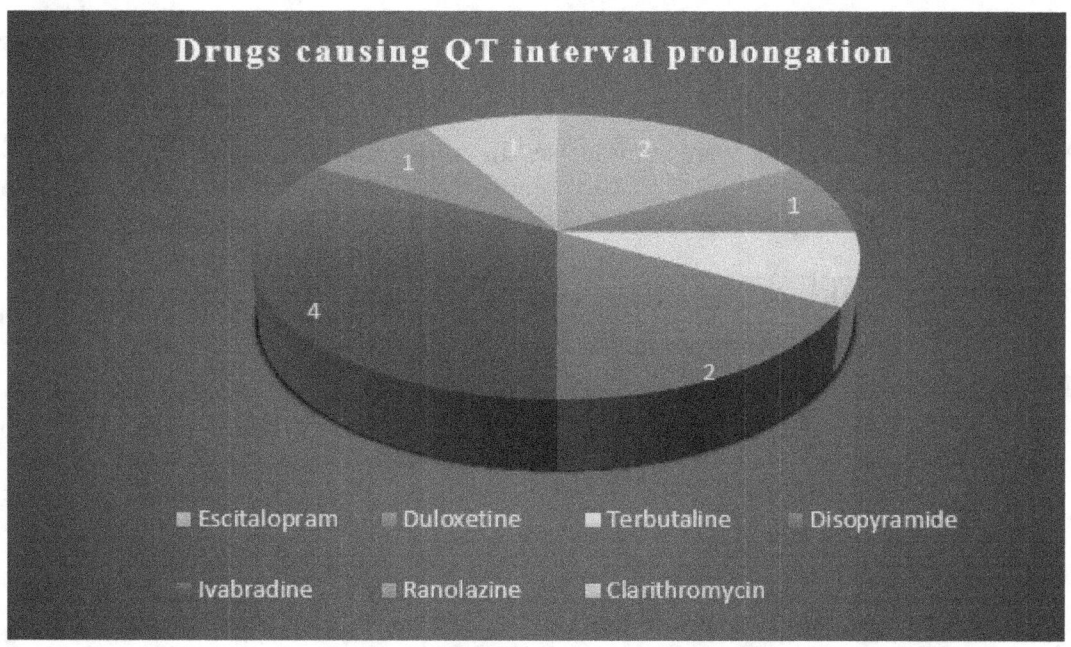

**Figure No- 7.15: Drugs Causing QT Prolongation**

Figure 7.16 and Table 7.8 summarises the percentage of the proportion of drugs causing QT interval prolongation.

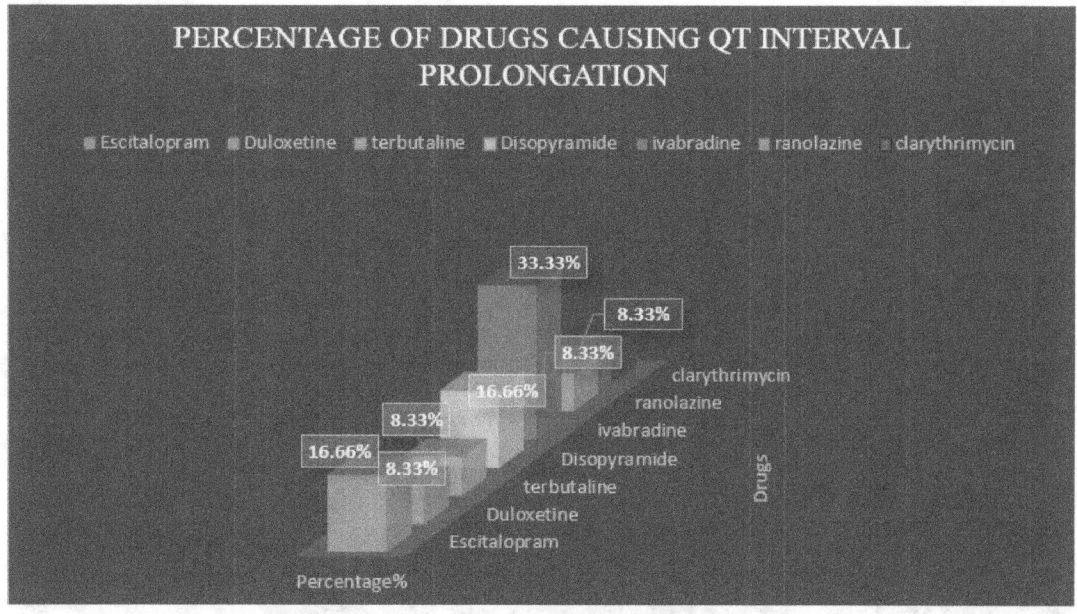

**Figure No- 7.16: Percentage of Drugs Causing QT Prolongation**

## 7.7 Drug Interactions Causing QT Prolongation:

The study discloses the list of combination of drugs that have known to cause QT interval prolongation in the cohort, out of which amiodarone and ranolazine have caused the highest drug interactions, which are summarised in table 7.9 and figure 7.17 below.

**Table No- 7.9: Drug Interactions Causing QT Interval Prolongation**

| S. No. | Drug-Drug interactions | Number (n) | Percentage % |
|--------|------------------------|-----------|--------------|
| 1. | amiodarone + ranolazine | 3 | 20% |
| 2. | sotalol + ranolazine | 1 | 6.60% |
| 3. | ivabradine + clarithromycin | 2 | 13.30% |
| 4. | escitalopram + amiodarone | 2 | 13.30% |
| 5. | clarythromycin + amiodarone | 1 | 6.60% |
| 6. | domperidone + escitalopram | 1 | 6.60% |
| 7. | domperidone + duloxetine | 1 | 6.60% |
| 8. | levofloxacin + voriconazole | 1 | 6.60% |
| 9. | ivabradine + imipramine | 1 | 6.60% |
| 10. | fluconazole + ranolazine | 1 | 6.60% |
| 11. | metoclopramide +levosulpride | 1 | 6.60% |
| | Total | 15 | 100% |

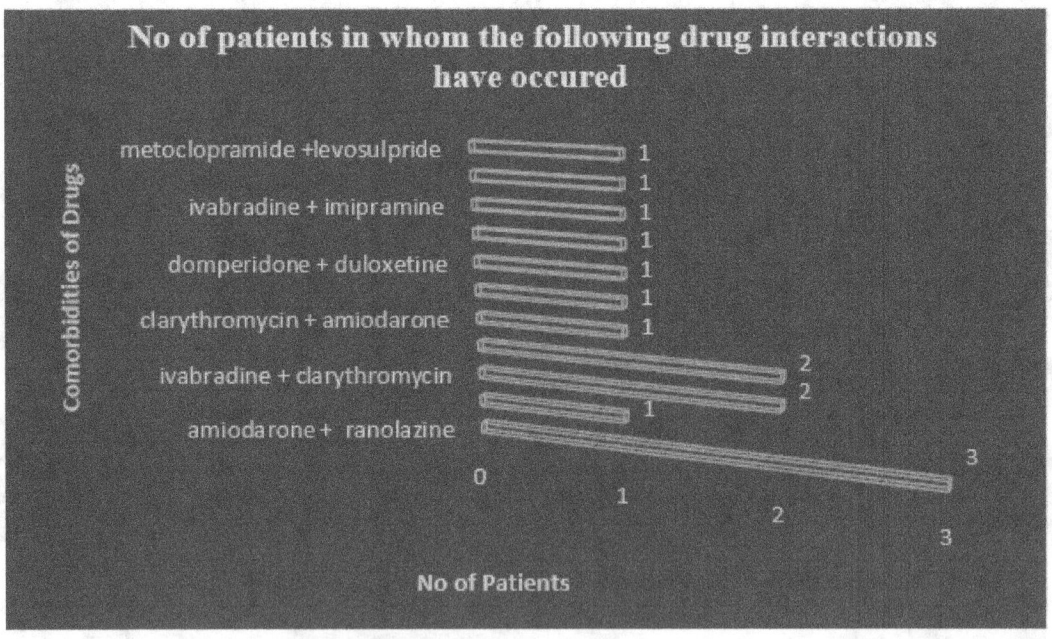

**Figure No- 7.17: Drug Interactions Causing QT Interval Prolongation**

The percentage of the proportion of the combination of drugs causing QT prolongation was accordingly calculated in table 7.9 and shown in figure 7.18.

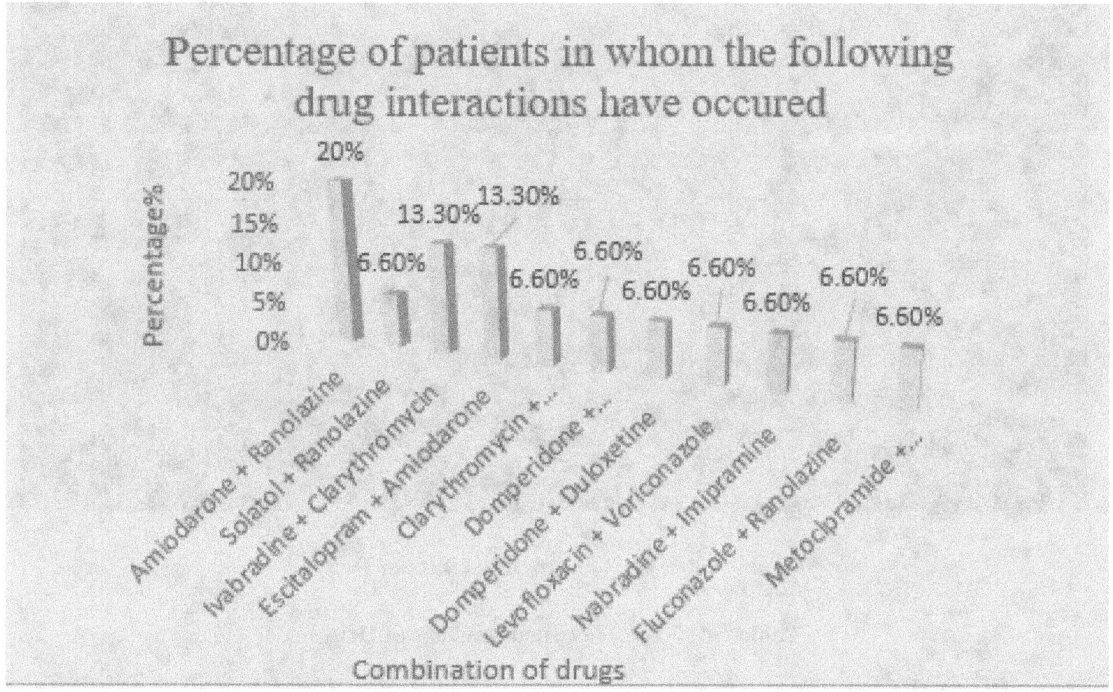

**Figure No- 7.18: Percentage of Drug Interactions Causing QT Interval Prolongation**

## 7.8 Therapeutic Class of Drugs:

This study categorises the drugs causing the QT interval prolongation into their therapeutic classes which are summarised in the table 7.10 and in the figure 7.19 that shows the highest therapeutic class of drugs that have caused long QT I.e. cardiovascular drugs.

Apart from the cardiac drugs the non-cardiac therapeutic class that has majorly caused QT prolongation are prokinetic agents and azole antifungals.

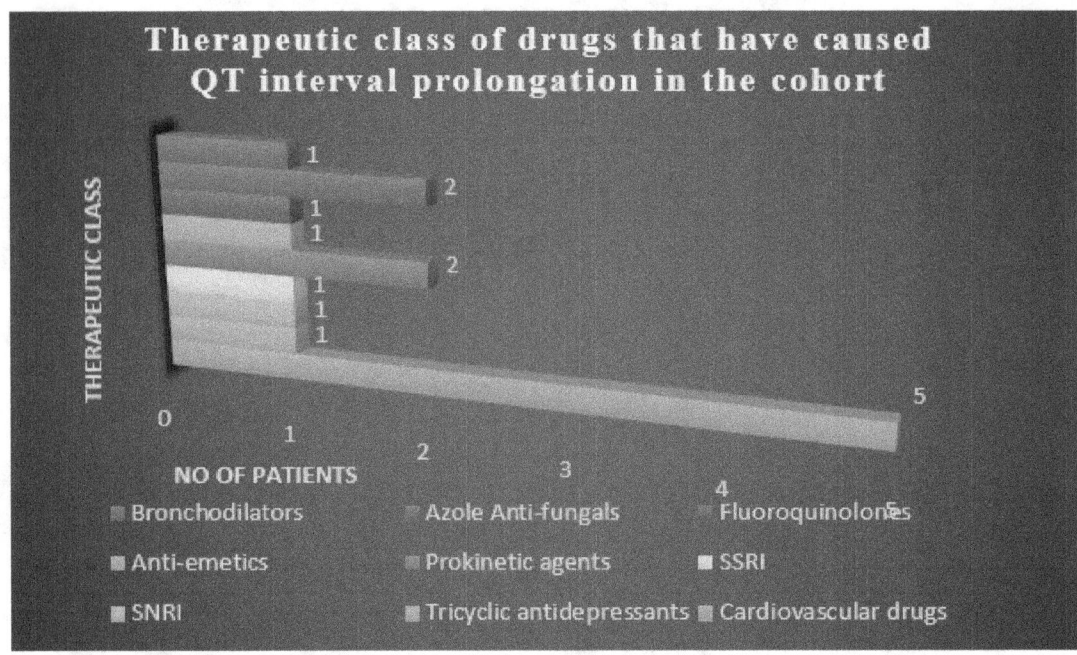

**Figure No- 7.19: Therapeutic Class of Drugs**

**Table No- 7.10: Therapeutic Class of Drugs**

| S. No | Therapeutic Class | Number (n) | Percentage % |
|-------|-------------------|------------|--------------|
| 1. | Cardiovascular drugs | 5 | 33.33% |
| 2. | Tricyclic antidepressants | 1 | 6.60% |
| 3. | SNRI (serotonin-norepinephrine reuptake inhibitors) | 1 | 6.60% |
| 4. | SSRI ( selective serotonin reuptake inhibitors) | 1 | 6.60% |
| 5. | Prokinetic agents | 2 | 13.30% |
| 6. | Anti-emetics | 1 | 6.60% |
| 7. | Fluoroquinolones | 1 | 6.60% |
| 8. | Azole Antifungals | 2 | 13.30% |
| 9. | Bronchodilators | 1 | 6.60% |
| | Total | 15 | 100% |

The percentage of therapeutic class of drugs involved in QT interval prolongation was calculated according and is summarised in table 7.10 and figure 7.20.

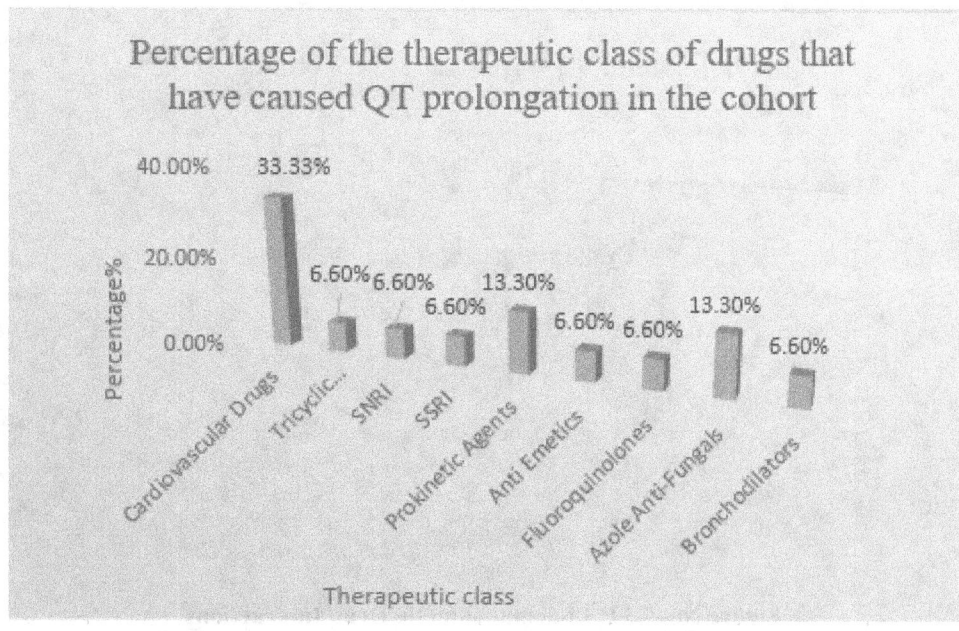

**Figure No- 7.20: Percentage of Therapeutic Class of Drugs**

## 7.9 Mechanism of Drug Interactions:

There are two major ways through which drug interactions occur which are pharmacokinetic and pharmacodynamic drug interactions in the founded QT prolonged cases. The drug combinations which have caused QT interval prolongation in the cohort population have been categorised with pharmacokinetic drug interactions and pharmacodynamic drug interactions which are enlisted in the table 7.11 and 7.12 respectively.

**Table No- 7.11: Pharmacokinetic Drug Interactions**

| S. No. | Pharmacokinetic Interactions | Number (n) | Percentage % |
|--------|------------------------------|------------|--------------|
| 1. | Clarithromycin +ranolazine | 1 | 25% |
| 2. | Clarithromycin +ivabradine | 1 | 25% |
| 3. | Clarithromycin +amiodarone | 1 | 25% |
| 4. | Fluconazole +ranolazine | 1 | 25% |
| | Total | 4 | 100% |

The graph below shows the proportion of drug combinations believed to have caused pharmacokinetic drug interactions in equal proportions which are shown in figure 7.21.

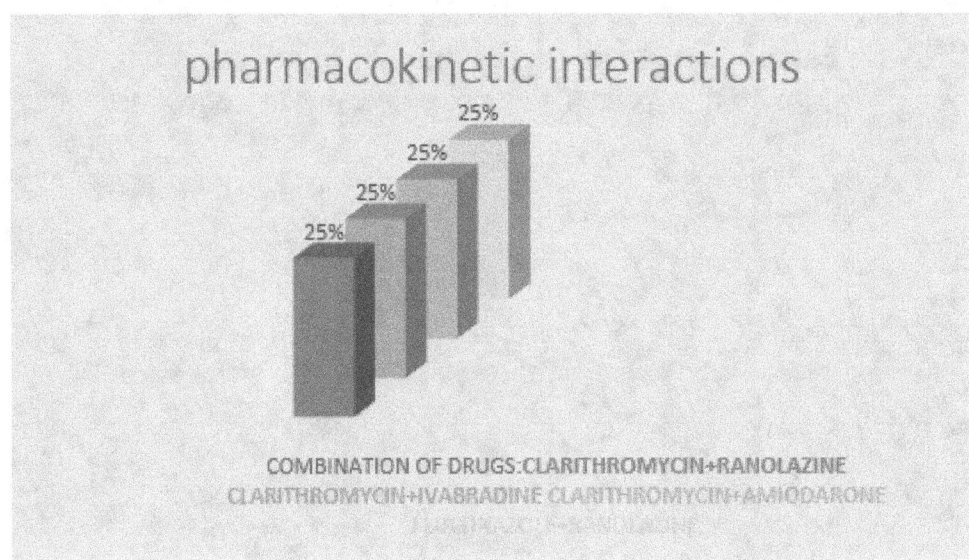

**Figure No- 7.21: Pharmacokinetic Drug Interactions**

Pharmacokinetic drug interactions were identified as occurred due to variation in the Cytochrome P450 enzyme-mediated metabolism of the drugs having the potential to cause QT interval prolongation. Drug interactions have been identified using Micromedex drug solutions. Potent CYP3A4 inhibitors such as clarithromycin and fluconazole which we have encountered in our study lead to an increase in the metabolism of drugs such as amiodarone, ivabradine, ranolazine and resulted in their longer half-life.

**Fluconazole with ranolazine:** concomitant use of fluconazole which is a potent inhibitor of CYP3A4 enzyme when given with a drug such as ranolazine, leads to an increased half-life of the drug in the blood of patients as a result, it will be more prolonged and can lead to serious QT interval prolongation. Fluconazole mediated CYP3A4 inhibition may continue for 4-5 days after discontinuation because of longer halflife[79].

**Ranolazine with clarithromycin:** Ranolazine and clarithromycin were given concomitantly, where clarithromycin can increase the ranolazine plasma levels by inhibition of cytochrome P450-3A mechanism and lead to QT interval prolongation and has an increased risk of cardiotoxicity.

**Amiodarone with clarithromycin:** Co-administration of amiodarone and clarithromycin should be avoided as increased plasma levels due to CYP3A4 inhibition by clarithromycin can cause longer half-life of amiodarone, in this case, the reaction can be sustained even after the discontinuation of the drug due to longer half-life of amiodarone which is 28-107 days[79].

**Ranolazine with fluconazole:** Even in this case fluconazole increases the plasma levels of ranolazine which can lead to QT interval prolongation.

Even though the combination of drugs such as clarithromycin &ranolazine and clarithromycin and ivabradine is contraindicated, they have been prescribed in the OPD settings[79].

Providentially, none of the patients has encountered the new kind of arrhythmia known as Torsades de pointes as a result of QT interval prolongation. The pharmacodynamic interactions have been listed in this study, out of which the drug combination of amiodarone and ranolazine have caused the highest number of interactions, which are summarised in table 7.12.

**Table No- 7.12: Pharmacodynamic Drug Interactions**

| S. No. | Pharmacodynamic Interactions | N | Percentage (%) |
|--------|------------------------------|---|----------------|
| 1. | Ivabradine +ranolazine | 1 | 8.33 |
| 2. | Ivabradine +imipramine | 1 | 8.33 |
| 3. | Sotalol +ranolazine | 1 | 8.33 |
| 4. | Amiodarone +escitalopram | 2 | 16.66 |
| 5. | Amiodarone +ranolazine | 3 | 25% |
| 6. | Levofloxacin +voriconazole | 1 | 8.33 |
| 7. | Domperidone +escitalopram | 1 | 8.33 |
| 8. | Domperidone +duloxetine | 1 | 8.33 |
| 9. | Metoclopramide +levosulpride | 1 | 8.33 |
| | Total | 12 | 100% |

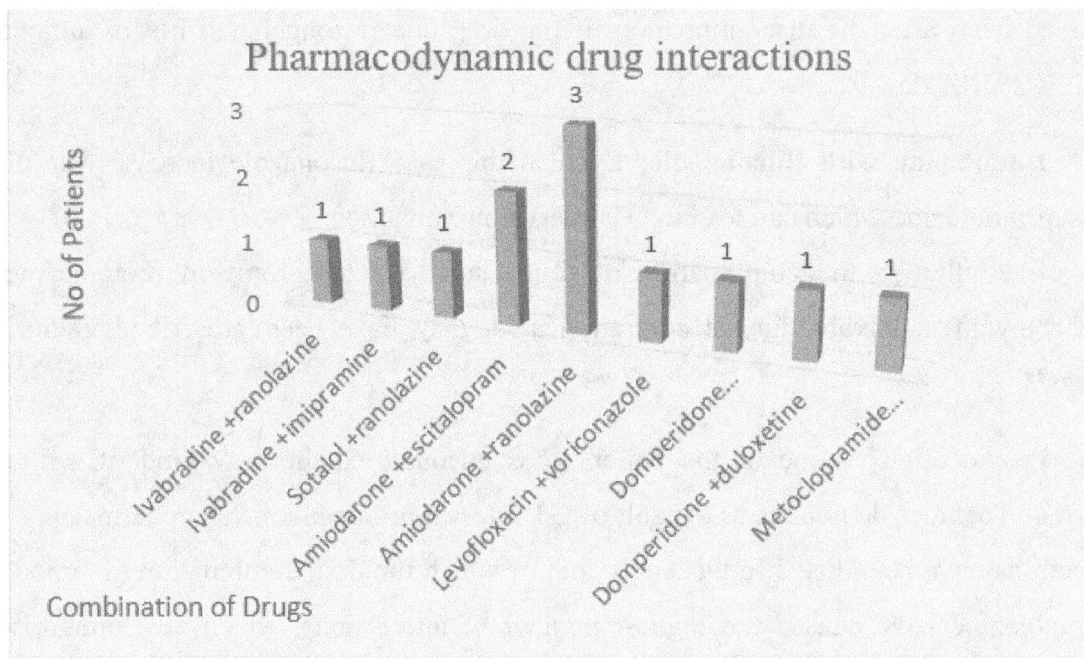

**Figure No- 7.22: Pharmacodynamic Drug Interactions**

The pharmacodynamic interactions have identified as occurred due to synergistic effects of two potentially QT-prolonging drugs.

The number of patients observed interactions and their percentages were calculated accordingly and have been shown in figure 7.22 and 7.23 respectively.

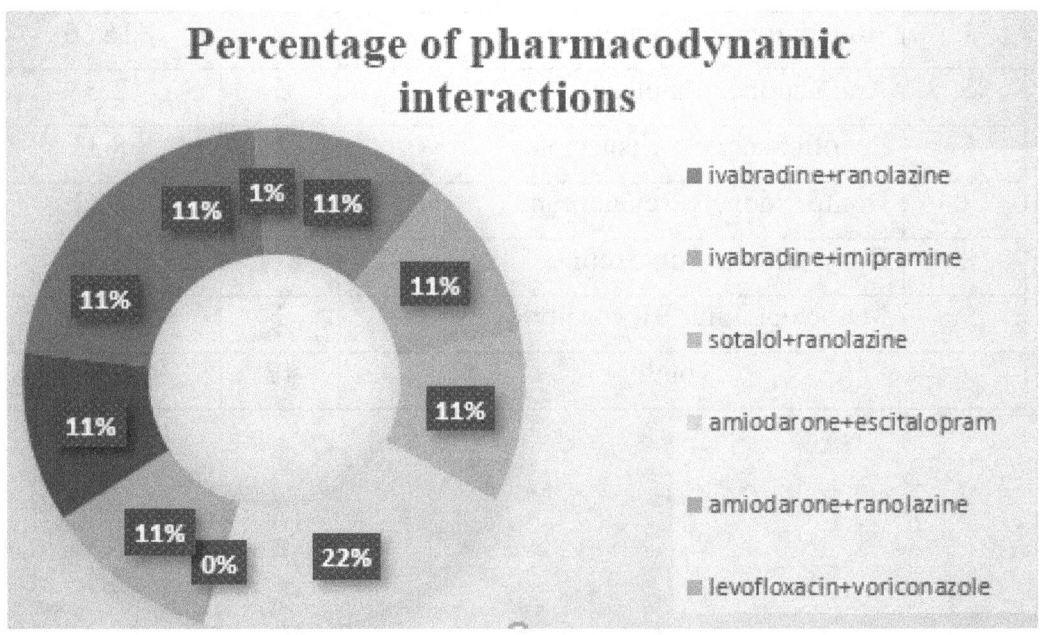

**Figure No- 7.23: Percentage of Pharmacodynamic Drug Interactions**

## 7.10 Cardiac and Non-Cardiac Drugs:

In the performed study the cardiac and non-cardiac drugs causing the QT interval prolongation were separated, out of which non-cardiac drugs were observed to have majorly caused the QT interval prolongation, which is summarised in the table 7.13 and figure 7.24.

**Table No- 7.13: Cardiac and Non-cardiac Drugs**

| S. No | Category | Number (n) | Percentage % |
|-------|----------|------------|--------------|
| 1. | Cardiovascular Drugs | 5 | 31% |
| 2. | Non-cardiac Drugs | 11 | 69% |
| | Total | 16 | 100% |

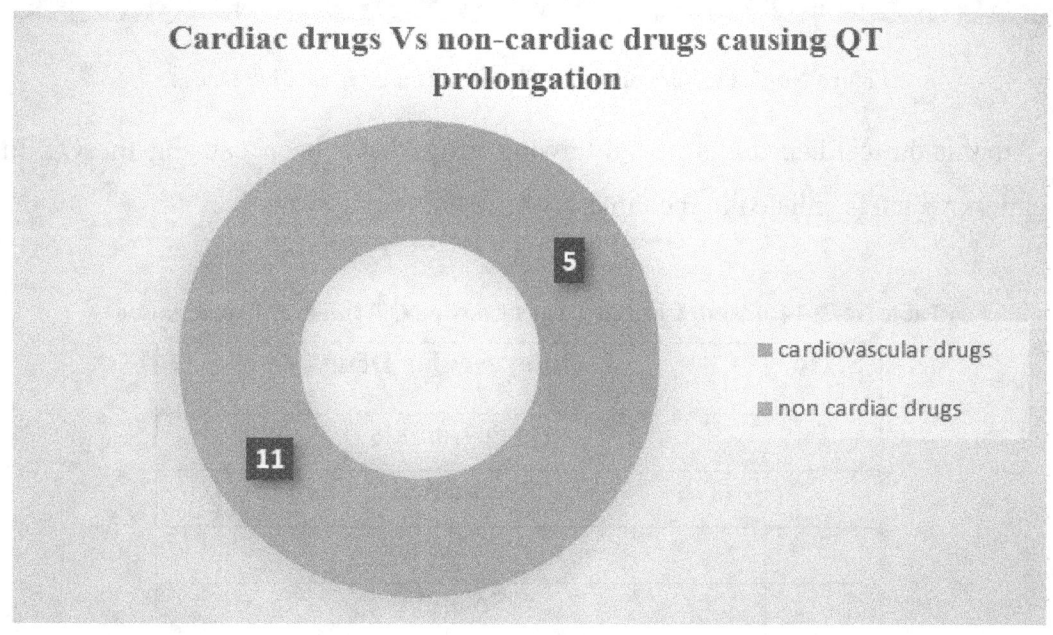

**Figure No- 7.24: Cardiac and Non-cardiac Drugs**

The percentages were calculated accordingly and are summarised in figure- 7.25

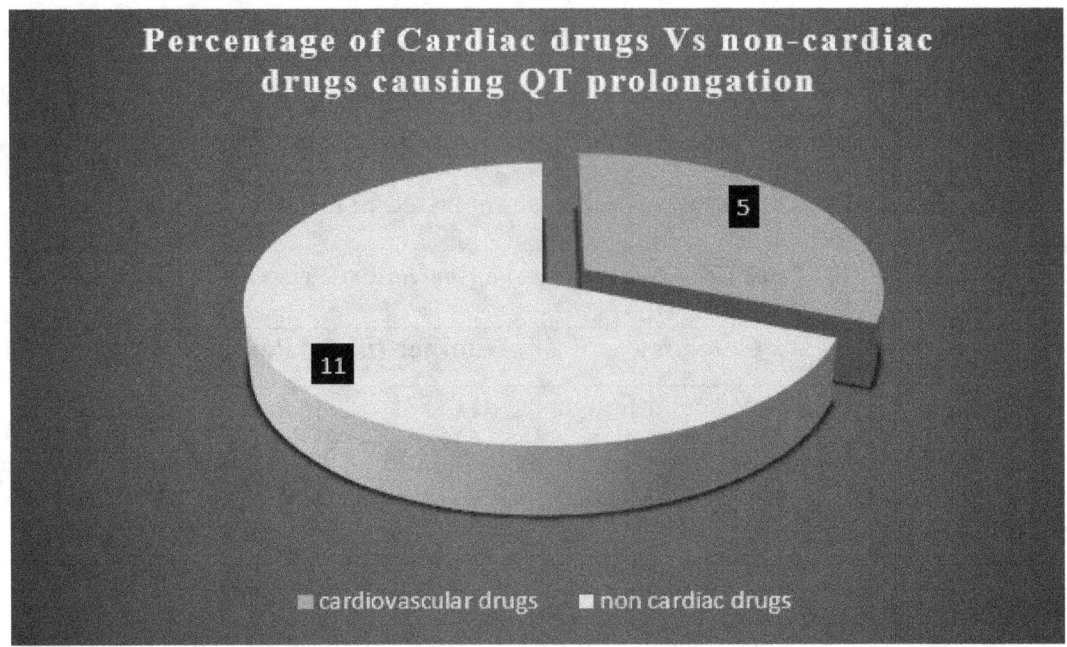

**Figure No- 7.25: Percentage of Cardiac and Non-cardiac Drugs**

Among the cardiac drugs, the following drugs have been causing the QT interval prolongation, which is enlisted in the table 7.14.

**Table No- 7.14: List of Cardiac Drugs Causing QT Interval Prolongation**

| S. No. | Cardiovascular Drugs |
|--------|----------------------|
| 1 | Disopyramide |
| 2 | Ivabradine |
| 3 | Ranolazine |
| 4 | Sotalol |
| 5 | Amiodarone |

Among the non-cardiac drugs, the following drugs have been causing the QT interval prolongation, which is listed in the table: 7.15.

**Table No- 7.15: List of Non-Cardiac Drugs Causing QT Interval Prolongation**

| S. No. | Non-cardiovascular Drugs |
|--------|--------------------------|
| 1 | Terbutaline |
| 2 | Escitalopram |
| 3 | Clarithromycin |
| 4 | Levofloxacin |
| 5 | Voriconazole |
| 6 | Imipramine |
| 7 | Fluconazole |
| 8 | Metoclopramide |
| 9 | Levosulpride |
| 10 | Domperidone |
| 11 | Duloxetine |

## 7.11 Standard Deviation:

The mean standard deviation values have been calculated for the QTc values after administration of the drug and after its withdrawal as QTc1 and QTc2 values and similarly, heart rate values as HR1 and HR2 respectively which have been summarised in the table 7.16.

**Table No- 7.16: Mean and Standard Deviation Values**

| S. No. | Parameter | Mean $\pm$ Standard deviation |
|--------|-----------|-------------------------------|
| 1 | QTc1 | $545.55 \pm 54.80$ |
| 2 | QTc2 | $490.84 \pm 45.85$ |
| 3 | HR1 | $84.51 \pm 16$ |
| 4 | HR2 | $83.85 \pm 20.75$ |

F-test was applied for the dual variables of QT interval after administration and withdrawal of drug, QTc1 and QTc2 and heart rate HR1 and HR2 which are summarised in the table: 7.17 and 7.18 respectively.

**Table No- 7.17: F-test for QTc 1 and QTc2 Variables**

| F-Test Two-Sample for Variances | | |
|---|---|---|
| Test | *Variable 1* | *Variable 2* |
| Mean | 545.5555556 | 490.4444444 |
| Variance | 3003.948718 | 2103.025641 |
| Observations | 27 | 27 |
| Df | 26 | 26 |
| F | 1.428393767 | |
| P(F<=f) one-tail | 0.184595958 | |
| F Critical one-tail | 1.929212675 | |

**Table No- 7.18: F-test for Heart Rate HR1 and HR2 Variables**

| F-Test Two-Sample for Variances | | |
|---|---|---|
| Test | *Variable 1* | *Variable 2* |
| Mean | 84.51852 | 83.85185 |
| Variance | 256.1054 | 430.9003 |
| Observations | 27 | 27 |
| Df | 26 | 26 |
| F | 0.59435 | |
| P(F<=f) one-tail | 0.095681 | |
| F Critical one-tail | 0.518346 | |

The study diagrammatically represents the mean deviation of QTc values from QTc1 to QTc 2 after the withdrawal of drug after QT prolongation due to a suspected drug. A linear correlation between the QTc1 and QTc2 values which shows a positive correlation between them was found. This is summarised in table 7.19 and a scatter plot in figure 7.26.

**Table No-7.19: Variables of QTc1 and QTc2**

| S. No. | QTc1 | QTc2 | S. No. | QTc1 | QTc2 |
|--------|------|------|--------|------|------|
| 1. | 527 | 499 | 15. | 555 | 464 |
| 2. | 511 | 489 | 16. | 545 | 486 |
| 3. | 542 | 439 | 17. | 546 | 490 |
| 4. | 502 | 483 | 18. | 555 | 490 |
| 5. | 701 | 589 | 19. | 497 | 488 |
| 6. | 519 | 511 | 20. | 687 | 588 |
| 7. | 551 | 502 | 21. | 616 | 478 |
| 8. | 505 | 470 | 22. | 447 | 531 |
| 9. | 543 | 510 | 23. | 450 | 536 |
| 10. | 538 | 491 | 24. | 540 | 481 |
| 11. | 529 | 499 | 25. | 562 | 475 |
| 12. | 568 | 513 | 26. | 568 | 498 |
| 13. | 535 | 410 | 27. | 544 | 352 |
| 14. | 547 | 480 | | | |

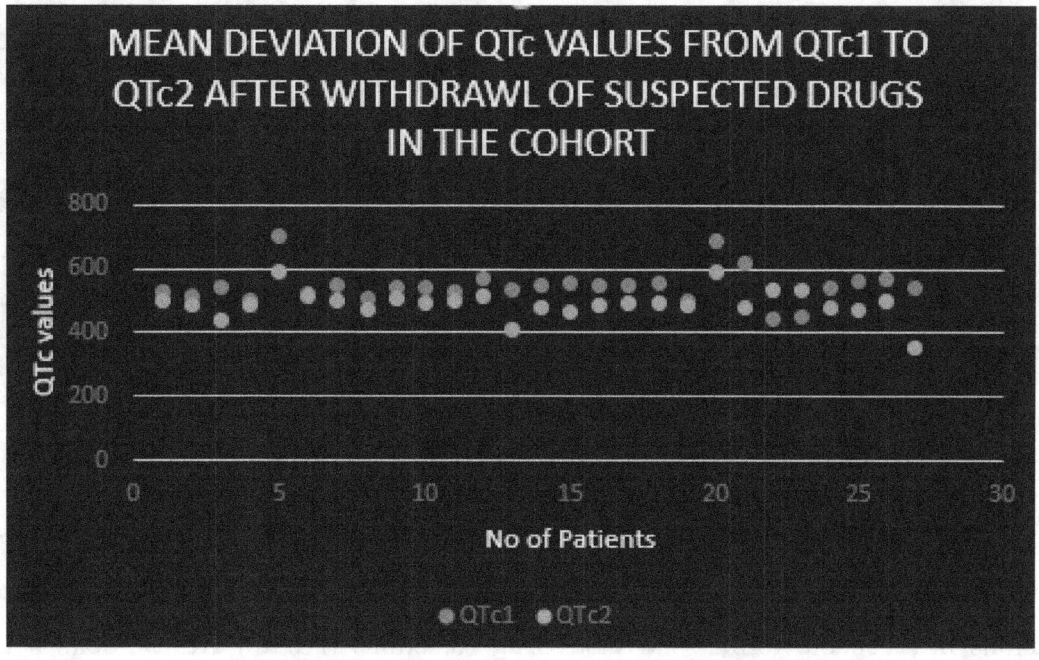

**Figure no-7.26: Mean Deviation of QTc Values from QTc1 to QTc2**

## 7.12 Causality Assessment:

In this study assessment of the causality of the drug-induced QT prolongation cases (n=12) by using the Naranjo scale of causality assessment and giving scores to each questionnaire. Out of which 9 cases were assessed to be probable and 3 cases definite which we have summarized in table 7.16 and figure 7.27.

Table No- 7.20: Causality Assessment Scoring

| S. No | Parameter | Scoring |
|-------|-----------|---------|
| 1. | Doubtful | 0 |
| 2. | Possible | 0 |
| 3. | Probable | 9 |
| 4. | Definite | 3 |

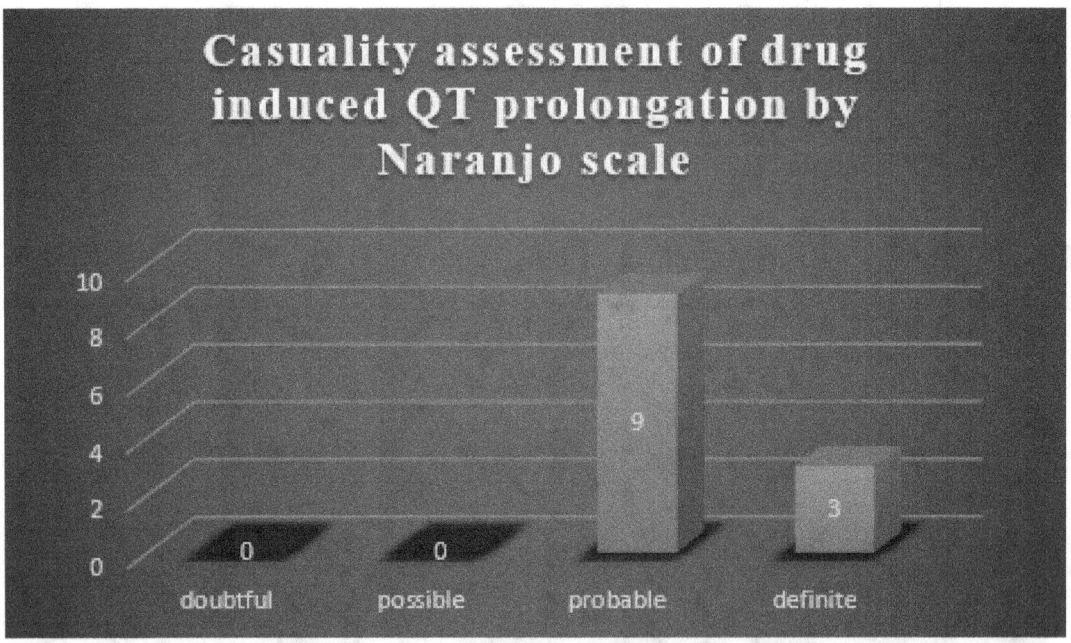

Figure No- 7.27: Causality Assessment Scoring

Variations in individual drug reactions will exist, so Naranjo's causality assessment scale has been used to assess the probability of a causal relationship for potential drug interactions in a reliable format. Adverse Drug Reactions (ADRs) are a crucial source of morbidities, so there is a need to form a causal association between the adverse event and the drugs[77]. The Naranjo causality assessment scale has been famous among physicians because of its simplicity[78].

The Naranjo scale of ADR probability was assessed by giving scores as -1 to +2 points for 10 questionnaires regarding the previous conclusive reports of the ADR, time of onset of reaction, recovery on stopping the drug, recurrence of reaction on challenging the drug, surrogate causes of the reaction, reappearance of reaction when placebo was given and assessed as scores $\geq 9$ as definite, 5-8 as probable, 1-4 as possible and 0 as doubtful[77]. The severity of all the reactions was assessed using the Karch and lasagne scale which states that the ADRs were moderate based on the fact that the patient required a change in the drug therapy.

## 7.13 Risk Factors:

The study enlists the risk factors present in patients who have experienced drug-induced and drug interactions causing QT interval prolongation out of which LV dysfunction has been the most frequently present risk factor in the patients which has been summarised in table 7.21and figure 7.28.

Table No- 7.21: Risk Factors

| S. No | Risk factors | No. of Patients | Percentage % |
|-------|--------------|-----------------|--------------|
| 1. | Bradycardia | 2 | 5.12% |
| 2. | Hypoglycaemia | 2 | 5.12% |
| 3. | Ischemic/dilated Cardiomyopathy | 7 | 17.94% |
| 4. | CHF/ADHF | 5 | 12.82% |
| 5. | LV dysfunction | 11 | 28.20% |
| 6. | hypertrophic cardiomyopathy | 5 | 12.82% |
| 7. | Valvular heart disease | 2 | 5.12% |
| 8. | Rheumatic heart disease | 2 | 5.12% |
| 9. | Hypothyroid | 2 | 5.12% |
| 10. | ischemic heart disease | 1 | 2.56% |
| | Total | 39 | 100% |

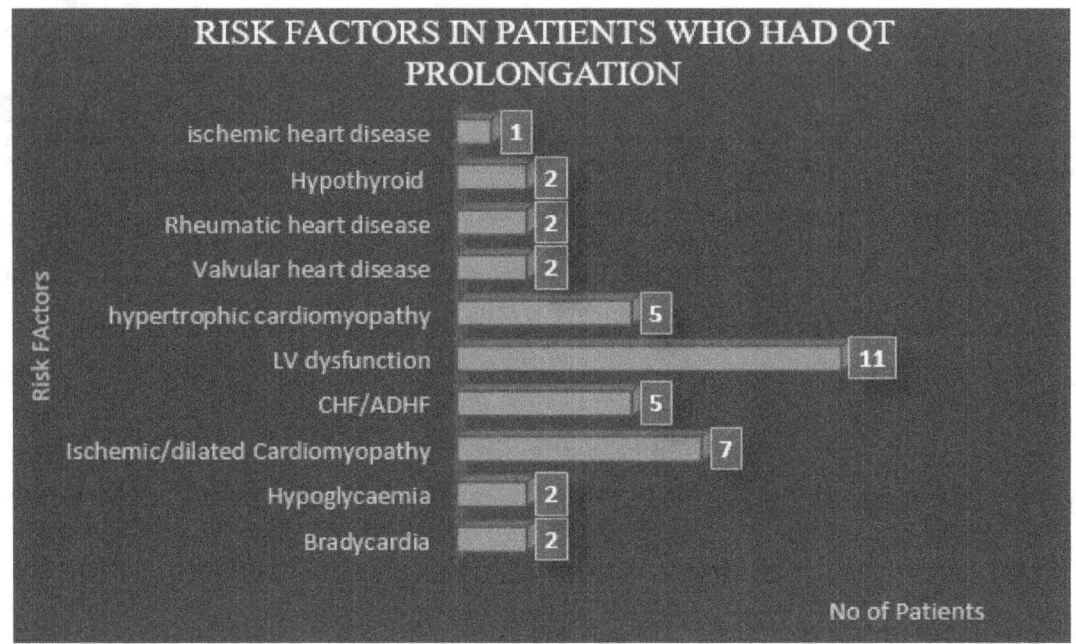

**Figure No- 7.28: Risk Factors**

The percentages of these prevalent risk factors were calculated accordingly and are shown in figure 7.29.

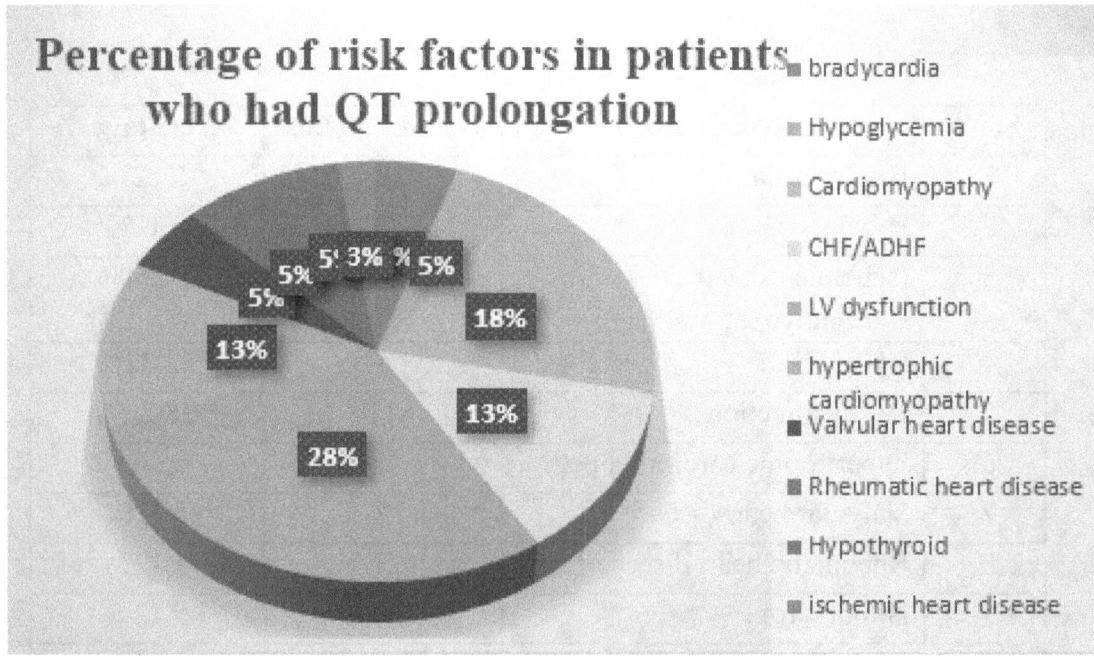

**Figure No- 7.29: Percentage of Risk Factors**

Risk factors in the patients were noted which play a significant role in the QT prolongation. The most frequent risk factor present was Left Ventricular dysfunction, then ischemic/dilated cardiomyopathy and hypertrophic cardiomyopathy in the lead.

From the total cases N=1122, 27 cases of QT interval prolongation were assessed for level of significance, $P$-value (probability value) which was found to be $P=0.08$.

# Chapter-8

# Summary

Prolongation of QT interval on ECG leads to life-threatening arrhythmia known as Torsades de pointes and ventricular fibrillation. It characterizes the total duration of ventricular electrical activity I.e. the time it takes for the heart ventricles to depolarize and repolarize.

It is measured on the 12 lead ECG and it is the primary diagnostic criteria in detection the QT interval prolongation. The normal duration of the QT interval should be in the range of 0.35 to 0.43 seconds. The QT interval needs to be corrected for what it would conceptually be at a heart rate of 60 beats per minute. The most frequently used formula is Bazzets formula.

There are two causes of QT interval prolongation one is congenital and the other acquired. The most prevalent cause of acquired long QT interval is drugs, though there are other causes such as electrolyte imbalance, bradycardia, heart diseases etc.

Before prescribing the drugs that prolong QT screening for risk factors should be done. Do not advise the patient for a QT-prolonging drug if he is already on antiarrhythmic agents. Baseline ECG must be performed in high-risk patients. Alternative drugs should be prescribed accordingly.

Drugs lead to QT interval prolongation by blocking the potassium channels thereby delaying action potential.

Some drugs affect the metabolism of other drugs which cause QT prolongation by affecting their metabolism. For example, drugs that inhibit CYP3A4 enzyme-like macrolides (clarithromycin, azithromycin, erythromycin) and antifungals (fluconazole) and are listed under pharmacokinetic drug interactions.

Concomitant use of more than one QT-prolonging drug leads to pharmacodynamic interaction.

In the present study, the prolonged QT interval cases were almost founded with 4% (n=47) and normal QT interval cases were of 96% (n=1075). From that QT prolonged cases, the male n=31(66%) population was more prevalent than the female n=16(34%) population among various causes of QT interval prolongation.

From this study of the total QT prolonged cases some specific drugs like ivabradine, escitalopram, domperidone, duloxetine, disopyramide, ranolazine, amiodarone etc, are majorly causing QT interval prolongation. Due to drug-induced QT Prolongation, the male n=18(67%) patients were more predominant than the female n=9 (33%) patients. An age interval group between 51-60 (n=9) were prone to suffer from QT prolongation as according to this present study and an age interval group between were 41-50 (n=1) were less prone to suffer from QT prolongation.

According to this study, we founded 5 major causes that lead to QT interval prolongation which includes drugs, drug-drug interactions, QRS complex widening, electrolyte imbalance and bradycardia. The study also estimated that ivabradine is the only drug which is more prevalent to cause QT interval prolongation.

This study also makes us understand that QT interval prolongation occurs due to pharmacodynamic and pharmacokinetic interactions based on their mechanisms. The study gives information there are of pharmacokinetic drug interactions n=4 and pharmacodynamic drug interactions are of n=9.

The study also reveals that QT interval prolongation occurs in almost all therapeutic classes of drugs (like tricyclic antidepressants, anti-emetics, prokinetic agents, antifungals, cardiovascular drugs etc) among which cardiac drugs were majorly causing the prolongation. Non-cardiac drugs also caused QT interval prolongation such as prokinetic agents (2) azole antifungals (2).

The mean deviation values were calculated for QTc1 and QTc2, HR1 and HR2 and resulted information was shown in the table.

An assessment of the causality of drug-induced QT interval prolongation cases n=12 was done by using the Naranjo scale of causality assessment and its scores were given based on a questionnaire. Out of which 9 cases were assessed to be as probable and 3 cases as definite.

In this prospective study QT, prolonged cases were also identified with risk factors (LV dysfunction, bradycardia, hypoglycaemia, hypertrophic cardiomyopathy etc) as this will be also considered as a causative factor to prolong QT interval. Out of the all these risk factors LV dysfunction was the most frequently reported risk factor from the total QT prolonged patients. The drug interactions and the adverse drug reactions have been identified using the Micromedex drug solutions and prospective observation respectively.

Assessment of Drug Interactions & Drugs Resulting in QT Interval Prolongation

# Chapter-9

# Conclusion

After analysing this study, it was concluded that QT interval prolongation needs regular monitoring of ECG before and after administration of drugs causing QT prolongation and should be corrected at the right time with already known databases. Alertness is more important concerning side effects and hazards during polypharmacy. Clinically QT-prolonging drugs should be avoided in already existing heart disease patients. The national drug formulary and committee for proprietary medicinal products (CPMP) gives information and its guidelines which can be used as a powerful indicator that reveals the issues of QT prolongation which will be helpful for careful examination. Additional research work is essential to estimate the potential QT prolongation.

# Chapter-10

# Future Outlook

1. The most frequently used formula in adjusting QT prolongation is Bazzett's formula however this method overcorrects QT interval at higher heart rates and under corrects it at very low heart rates, to avoid this issue other correction formulas such as Fridericia and Framingham formulas should be used.

2. The Naranjo scale was utilised to help standardize assessment of causality for all adverse drug reactions but it has less sensitivity and specificity in assigning causality, hence in future studies, other scales like WHO-UMC (World health organisation-Uppsala monitoring centre) causality assessment scale.

3. For more accuracy of identified drug interactions, DIPS (Drug interaction probability assessment scale) score can also be used for assessing drug interactions in the future related studies.

4. Population attributable risk (PAR) can also be calculated in the future study as well as odds ratio using regression analysis can be used.

Future Outlook

# Chapter-11
# Limitations

➢ This is a single centred study.

➢ This study has only focused on cardiac outpatient department which is especially an electrophysiology department, whereas QT prolongation can also be seen in other departments such as neurology, pulmonology etc.

➢ This study has excluded Inpatient cases, where there is more scope of QT interval prolongation cases in hospitalized patients.

➢ This study has excluded age group below 18 yrs.

Limitations

Assessment of Drug Interactions & Drugs Resulting in QT Interval Prolongation

# References

1. Atul Luthra. Abnormalities of QT interval. Textbook of ECG Made Easy. The Health Sciences Publisher, New Delhi. 5th edition. 2017. ISBN:978-93-86150-21-9:121.

2. Cox, Natalie K RN. The QT interval: How long is too long. Lippincott Williams & Wilkins. Nursing made incredibly easy. March-April 2011;9(2):17-21.

3. Taggart NW, Haglund CM, Tester DJ, Ackerman MJ. Diagnostic miscues in the congenital long QT syndrome. American Heart Association journal. 2007 May 22;115(20):2613-20.

4. Postema PG, De Jong JS, Van der Bilt IA, Wilde AA. Accurate electrocardiographic assessment of the QT interval: teach the tangent. Heart Rhythm Journal. 2008 July;5(7):1015-8.

5. Johnson JN, Ackerman MJ. QTc: How long is too long? British Journal of Sports Medicine. 2009 September;43(9):657-662.

6. Yee Guan Yap, A John Camm. Drug induced QT prolongation and torsades de pointes, BMJ Heart Journal. 2007 89(11):1363-1372.

7. Leo Schamroth. Basic Priciples. Cardiac channelopathies. An introduction to electrocardiography. eighth adapted edition. Wiley India Pvt. Ltd. 2013. ISBN:978-81-265-3897-3. 16-18, 202-203.

8. Moss AJ, Robinson JL. The Long QT syndrome: Genetic considerations, Journal of Cardiovascular Medicine. 1992; 2: 81-3.

9. Bert Vandenberk, Eline Vandael, Tomas Robyns, Joris Vandenberghe, Christophe Garweg, Veerle Foulon. Which QT correction formulae to use for QT monitoring. Journal of the American heart association. 2016;5:e003264.

10. Anand Ambhore, Swee Guan Teo, Abdul Razakjr Bin Omar, Kian-Keong Poh. ECG series. Importance of QT interval in practice. Singapore Medical Journal. 2014 Dec; 55(12): 607–612.

11. L. Brent Mitchell. Long QT syndrome and Torsade's de Pointes Ventricular Tachycardia. MSD MANUALS, September 2017.

12. Crotti L, Celano G, Dagradi F, Schwartz P. Congenital long QT syndrome. Orphanet Journal of Rare Diseases. July 2008 July 7; 3:18.

13. Atul Luthra. Abnormalities of QT interval. Textbook of ECG Made Easy, 5th edition: 2017, The Health Sciences Publisher, New Delhi, ISBN: 978-93-86150-21-9:123.

14. Atul Luthra. Abnormalities of QT interval. Textbook of ECG Made Easy,5th edition:2017, The Health Sciences Publisher, New Delhi, ISBN:978-93-86150-21-9:125.

15. Charles I Berul. Acquired long QT Syndrome. Jan 03, 2018. https://www.uptodate.com.

16. Heist EK, Ruskin JN. Drug-Induced proarrhythmic and use of QTc-prolonging agents: clues for clinicians. Journal of Heart Rhythm. 2005 Nov; 2(2 Suppl): S1-8.

17. Yee Guan Yap, A John Camm: Drug induced QT prolongation and torsade's de pointes, BMJ Journals year 2007. 89(11):1363-1372.

18. Viskin S. Long QT syndromes and torsade de Pointes. The Lancet Journals. 1999 November 6; 354(9190):1625-1633.

19. Morrissette P, Hreiche R, Turgeon J. Drug-Induced long QT syndrome and torsade's de Pointes. Canadian Journal of Cardiology. 2005;21(10):857-864.

20. Al-Khatib SM, La Pointe NM, Kramer JM, Califf RM et al. What clinicians should know about the QT interval. The Journal of American Medical Association. 2003 April 23-30; 289(16): 2120-2127.

21. Fraley MA, Birchem JA, Senkottaiyan N, Alpert MA et al. Obesity and the electrocardiogram. Obesety Review Journals. 2005 November; 6(4): 275-81.

22. Schwartz PJ, Woosley RL. Predicting the Unpredictable: Drug-Induced QT Prolongation and Torsade's de Pointes. Journal of American College of Cardiology. 2016 April 5; 67(13):1639-1650.

23. MEDSAFE Newzealand Medicines and Medical Devices Safety Authority. Drug-induced QT prolongation and Torsade's de Pointes – the jacks. December 2010.31(4):27-29.

24. Post Script Extra. NHS Greater Glasgow and cycle. December 2012.211. http://www.ggcprescribing.org.uk/media/uploads/ps_extra/pse_21.pdf.

25. Antzelevitch C, Sicouri S. Clinical relevance of cardiac arrhythmias generated by after depolarisation: role of M cells in the generation of U ware, triggered activity and torsade de pointes. Journal of American College of Cardiology 1994;23:259-277.

26. Curran ME, Splawski I, Timothy KW, Vincent GM,Green ED, Keating MT. A molecular basis for cardia arrhythmia: HERG mutations cause long QT syndrome. Cells Journal 80(5);795-803.

27. Sanguinetti MC, Keating MT et al: Role of delayed rectifier potassium channels in cardiac repolarisation and arrhythmias. News Physiological Sciences 12, 152-157.

28. Drew BJ, Ackerman MJ, Funk M, et al. Prevention of torsade de pointes in hospital settings: a scientific statement from the American Heart Association and the American College of Cardiology Foundation endorsed by the American Association of Critical-Care Nurses and the international. Society for Computerize Electro cardiology. Journal of American College of Cardiology. 2010;5:934-47.

29. Yang T, Chun YW, Stroud DM, et al. Screening for acute $I_{Kr}$ block is insufficient to detect torsades de pointes liability: role of late sodium current. Circulation Journal. 2014;130:224-34.

30. Roden DM. Torsades de pointes. Clinical Cardiology. 1993;16:683-6.

31. Sandra JG, Megan EM, Mei TL. Evaluation of the use of electrocardiogram monitoring in patients on psychotropic medications that have a risk of QT prolongation. Mental Health Clinician Journal. 2016;(4):171-177.

32. Khan IA. Long -QT syndrome: diagnosis and management. American Heart Journal. 2002;143:7-14.

33. Tzivoni D, Banai S, Schuger C. Treatment of torsades de pointes with magnesium sulphate. Circulation Journal. 1988;77:392.

34. Magnano AR, Holleran S, Ramakrishnan R, Reiffel JA, Bloomfield DM. Autonomic nervous system influences on QT interval in normal subjects. Journal of American College of Cardiology 2002; 39:1820-6.

35. Ramesh M Gowda, Ijaz A Khan, Sabrina LW, Balendu CV, Terrence JS. International Journal of Cardiology. April 2003, 96 (2004) 1-6.

36. Monraba R, Sala C. Percutaneous overdrive pacing in the out-of-hospital treatment of torsades de pointes. Annals of Emergency Medicine. 1999; 33:356-7.

37. Noda T, Takaki H, Kurita T, et al. Gene-specific response of dynamic ventricular repolarization to sympathetic stimulation in LQT1, LQT2 and LQT3 forms of congenital long QT syndrome. European Heart Journal. 2002; 23:975-83.

38. Dorostkar PC, Eldar M, Belhassen B, Scheinman MM. Long-term follow-up of patients with long-QT syndrome treated with beta-blockers and continuous pacing. Circulation Journal 1999; 100:2431-6.

39. Groh WJ, Silka MJ, Oliver RP, et al. Use of implantable cardioverter defibrillators in the congenital long-QT syndrome. American Journal of Cardiology. 1996; 78:703-6.

40. Compton SJ, Lux RL, Ramsey MR, et al. Genetically defined therapy of inherited long-QT syndrome. Correction of abnormal repolarization by potassium. Circulation Journal. 1996;94:1018-22.

41. Shimizu W, Antzelevitch C. Effects of a K (+) channel opener to reduce transmural dispersion of repolarization and prevent torsades de pointes in LQT1, LQT2 and LQT3 models of the long-QT syndrome. Circulation Journal. 2000;102:706-12.

42. Schwartz P, Priori S, Locati E, et al: Long QT syndrome patients with mutations of the SCN-and HERG genes have differential responses to NA channel blockade and to increases in heart rate. Circulation 1995;92:3381-6.

43. Benhorin J, Taub R, Goldmit M, et al. Effects of flecainide in patients with new SCN-mutation: mutation specific therapy for long-QT syndrome. Circulation Journal. 2000; 101:1698-706.

44. Pratt CM, Singh SN, Al-Khalidi HR, Brum JM, Holroyde MJ, Marcello SR, et al. The efficacy of azimilide in the treatment of atrial fibrillation in the presence of left ventricular systolic dysfunction: Results from the Azimilide Postinfarct Survival Evaluation (ALIVE) trial. Journal of American College of Cardiology. 2004;43(7):1211-6.

45. International Conference on Harmonisation: Guidance on E14 Clinical Evaluation of QT/QTc Interval Prolongation and Proarrhythmic Potential for Non-Antiarrhythmic Drugs availability. Food and Drug Administration, HHS.Fed Regist. 2005 Oct 20;70(202):61134-5.

46. Samarendra P1, Kumari S, Evans SJ, Sacchi TJ, Navarro V. QT prolongation associated with azithromycin/amiodarone combination. Journal of Pacing and electrophysiology. 2001;24(10):1572-1574.

47. FDA. Drug development and drug interactions. Table of substrates, inhibitors and inducers.http://www.fda.gov/Drugs/DevelopmentApprovalProcess/DevelopmentReso urces/DrugInteractionsLabelling/ucm093664.html. Accessed June 20, 2015.

48. Barbara winiowasala, Zofia Tylutki, Gabriela Wyszogrodzka, Sebastian Polak. Drug-drug interactions and QT prolongation as a commonly assessed cardiac effect comprehensive overview of clinical trials. BMC Journal pharmacology and toxicology. 2016; 17:12.

49. Who Health Organisation. Cisapride. Pharmaceuticals: Restrictions in use and availability. essential drugs and medicines-quality assurance and safety of medicines health technology and pharmaceuticals. 2001; EDM/QSM/2001.3.

50. Sarganas G, Garbe E, Klimpel A, et al. Epidemiology of symptomatic drug-induced long QT syndrome and Torsade de Pointes in Germany. Europace Journal. 2014, 16:101–108.

51. Boyce MJ, Baisley KJ, Warrington SJ. Pharmacokinetic interaction between domperidone and ketoconazole leads to QT prolongation in healthy volunteers: a randomized, placebo-controlled, double-blind, crossover study. British Journal of Clinical Pharmacology. 2012;73(3):411-421.

52. Hennekens CH, Buring JE. Epidemiology in Medicine. Philadelphia: Lippincott-Raven Publishers, 1987.

53. Beach SR, Celano CM, Noseworthy PA, Alan MSMB, Catlin Adams MD, et al. QTc prolongation, torsades de pointes, and psychotropic medications: A 5 year Update. Psychosomatics. 2013; 54:1-13.

54. Kounas SP, Letsas KP, Sideris A, Efraimidis M, Kardaras F. QT interval prolongation and torsades de pointes due to a coadministration of metronidazole and amiodarone. Journal of Pacing Clinical Electrophysiology 2005;28(5):472–473.

55. Woosley, RL, and Romero, KA. QT drugs List. AZCERT (Arizona Centre For Education and Research on Therapeutics). 13 September 2015. www.Crediblemeds.org.

56. Vlase L, Popa A, Neag M, Muntean D, Leucuta SE. Pharmacokinetic interactions study between ivabradine with fluoxetine or metronidazole in healthy volunteers. International Journal of Farmacia. 58, 471-477.

57. Iftikhar. Nelfinavir and Ivabradine Added to QTdrugs.org Lists April 17, 2014. http://crediblemeds.org/blog/nelfinavir-and-ivabradine-added-list-drugs-avoid-/

58. Roden DM. Drug induced prolongation of the QT prolongation. New England Journal of Medicine. 2004;350(10):1013-1022.

59. Roden DM, Woosley RL, Primm RK. Incidence and clinical features of the quinidine-associated long QT syndrome implications for patient care. American Heart Journal. 1986:1088–1093.

60. Yap YG, Camm AJ. Drug induced QT prolongation and torsades de pointes. American Heart Journal. 2003:1363–1372.

61. Sohaib SM, Papacosta O, Morris RW, Macfarlane PW, Whincup PH. Length of the QT interval determinants and prognostic implications in a population-based prospective study of older men. Journal of Electrocardiology. 2008:704–710.

62. AZCERT Inc. Combined list of drugs that prolong QT and/or cause Torsades de Pointes (TdP). 2017. https://crediblemeds.org/pdftemp/pdf/CombinedList.pdf.

63. Solai LK, Mulsant BH, Pollock BG. Selective serotonin reuptake inhibitors for late-life depression: a comparative review. Drugs and Aging. 2001;18(5):355–363.

64. Sauer AJ, Newton Cheh. Clinical and genetic determinants of torsade de pointes risk. Circulation Journal. 2012:1684–1694.

65. Fisher AA, Davis MW. Prolonged QT interval, syncope, and delirium with galantamine. The Annals of pharmacotherapy. 2008;42(2):278–283.

66. Bertino JS Jr, Owens RC Jr, Carnes TD, Iannini PB. Gatifloxacin associated corrected QT interval prolongation, torsades de pointes and ventricular fibrillation in patients with known risk factors. Infectious Disease Society of America. 2002;34(6):861–886.

67. Viskin S, Justo D, Halkin A, Zeltser D. Long QT syndrome caused by non-cardiac drugs. Progress in cardiovascular disease Journal. 2003;45(5):415–427.

68. Gilbert DN, Moellering RC Jr, Sande MA, Ed. Sanford Guide to Antimicrobial Therapy. Jeb C Sanford Publishers, Hyde Park VT. 2002.

69. Sorawicz B, Knoebel SB. Long QT: good, bad, or indifferent. Journal of American college of cardiology. 1984;4(2):398–413.

70. Haddad PM, Anderson IM. Antipsychotic-related QTc prolongation, torsade de pointes and sudden death. Drugs. 2002;62(11):1649–1671.

71. Haverkamp W, Breithardt G, Camm AJ, et al. The potential for QT prolongation and pro-arrhythmia by non-anti-arrhythmic drugs: clinical and regulatory implications. Report on a Policy Conference of the European Society of Cardiology. Journal of cardiovascular research. 2000;47(2):219–233.

72. Antonelli D, Atar S, Freedberg NA, et al. Torsade de pointes in patients on chronic amiodarone treatment: contributing factors and drug interactions. The Israel medical association journal. 2005;7(3):163-5.

73. Mörtl D, Agneter E, Krivanek P et al. Dual rate-dependent cardiac electrophysiologic effects of haloperidol: slowing of intraventricular conduction and lengthening of repolarization. Journal of cardiovascular pharmacology and therapeutics. 2003;41(6):870-9.

74. Zareba W, Lin DA. Antipsychotic drugs and QT interval prolongation. Psychiatric quarterly. 2003; 74(3):291-306.

75. Dabhi J, Mehtha A. QT prolongation on drug safety in Indian population. Current drug safety. 2007;2(3):200-203.

76. Chhagan Lal Birda, Ashish Bhalla, Naveet Sharma, Savitha Kumari. Prevalence and prognostic significance of prolonged QTc interval in emergency medical patients: A prospective observational. International Journal of Injury critical illness and science. 2018;8(1):28-35.

77. Syed Ahmed Zakim. Adverse drug reactions and casualty assessment scale drug India. Lung India Journal. 2011;28(2):152-153.

78. Naranjo CA, Busto U, Sellers EM, Sandol P. A method for estimating the probability of adverse drug reactions. Journal of clinical pharmacology and therapeutics. 1981;30(2):239-245.

79. Micromedex Drug Solutions. http://www.micromedex.com/clinicalknowledge.

References

Assessment of Drug Interactions & Drugs Resulting in QT Interval Prolongation

## STUDY PROFORMA

**Table No- 1: Patient Demographic Details**

| Patients ID No: | Department: | Age: | Gender: |
|---|---|---|---|
| Date of visit: | Height: | Weight: | BMI: |

**Chief complaints:**

**Past medical history:**

**Social history:**

Smoking: yes/no if yes_____packs/day

Chewing tobacco: yes/no if yes_____quantity

**Allergies:**

**Family history:**

**Table No- 2: Past Medication History:**

| S. No: | Drug | Dose | Route | Frequency | Duration | Indication |
|---|---|---|---|---|---|---|
|  |  |  |  |  |  |  |

**Table No- 3: Diagnosis**

| Clinical Diagnosis: |
|---|
| Cardiac Diagnosis: |

**Left ventricular ejection fraction (LVEF):**

|  |
|---|
|  |

**Table No- 4: Current Therapy:**

| S. No. | Drug | Dose | Route | Frequency | Duration | Indication |
|---|---|---|---|---|---|---|
|  |  |  |  |  |  |  |

**Table No- 5: ECG Parameters**

| HR(heart rate) | |
|---|---|
| PR interval | |
| QRS complex | |
| QT interval | |
| QTc interval | |

**Calculation of QTC by Using Bazzets Formula:**

QTc (Corrected QT interval) = $\dfrac{\text{QT Interval in Seconds}}{\text{Root RR in Seconds}}$

**Table No- 6: Prolonged QTc Interval**

| Past ECG | QTc & HR Day 1 | QTc & HR Day 3 |
|---|---|---|
| QTc:<br><br>HR: | QTc:<br><br>HR: | QTc:<br><br>HR: |

**Table No- 7: Potential Drug-Drug Interaction**

| S. No. | Drug | Dose | Route | Frequency | Start date | Stop date |
|---|---|---|---|---|---|---|
| | | | | | | |

**Table No- 8: Mechanism**

| Object Drug | Precipitant Drug | Pharmacokinetic Interaction | Pharmacodynamic Interaction | Mechanism/Effect |
|---|---|---|---|---|
|  |  |  |  |  |

**Table No- 9:  Management**

|  |
|---|
|  |

**Table No- 10: Diagnostic Tests**

| Diagnostic tests | Risk factors: |
|---|---|
| Potassium: | Bradycardia: |
| Calcium: | Female: <br> Obesity: |
| Magnesium: | Structural Heart Diseases : |
| T3: | Geriatric: |
| T4: | Hepatic/renal Impairment : |
| TSH: | Family History of Sudden Death: |

**Table No- 11: Others**

|  |
|---|
|  |

# INFORMED CONSENT FORM

Subject Identification number for this study. Title of the Project:

**"Assessment of Serious Drug Interactions and Drugs Resulting in QT Interval Prolongation in Cardiac Outpatient at a Tertiary Care Hospital"**

Name of the Guide:  Dr. P. Sagar                                    Tel. No: 8121898000

I have received the information sheet on the above study and have read and / or understood the written information.

I have been given the chance to discuss the study and ask questions. I understand that no new drugs would be used during the study and only routine laboratory tests are included in the study.

I consent to take part in the study and I am aware that my participation is voluntary. I understand that I may withdraw at any time without this affecting my future care.

I understand that the information collected about me from my participation in this study and sections of any medical notes may be looked at by responsible persons (ethics committee members/ regulatory authorities). I give access to these individuals to have access to my records. I understand that no new drug is being used in this study .Also, the study involves conduction of only routine laboratory tests.

_____                    _____

Name/Thumb Impression of Subject                    Date/Signature

----------------------------------------------                    -------------------------------

Name/Legally accepted representative (LAR)                    Date/Signature

_____                    _____

Name of the person obtaining consent                    Date/signature

తెలియజేయబడిన ఆంగీకార పత్రం

రోగి గుర్తింపు సంఖ్య: _____

ప్రాజెక్టు పేరు: తృతీయ కేర్ ఆసుపత్రిలో రోగికి QT విరామం పొడిగింపు ఫలితంగా తీవ్రమైన ఔషధ సంకర్షణలు మరియు ఔషధాల అంచనా.

గైడ్ పేరు: పి. సాగర్                         ఫోన్ నం. 8121898000

పైన తెలుపబడిన అధ్యయనంపై సమాచార పత్రాను నేను స్వీకరించాను మరియు చదివాను మరియు/లేదా వ్రాయబడిన సమాచారాన్ను అర్థం చేసుకున్నారను.

సదరు అధ్యయనంను చర్చించే మరియు ప్రశ్నలను అడిగే అవకాశం నాకు ఇవ్వబడినది. నేను అర్థం చేసుకున్నది ఏమనగా అధ్యయనంలో కొత్త మందులు ఉపయోగించబడవు మరియు యధావిధి ప్రయోగశాల పరిక్షలు మాత్రమే నిర్వహించబడును.

సదరు అధ్యయనంలో పాల్గొనటలకు నేను అంగీకరిస్తున్నునుమరియు నేను స్వచ్ఛందంగా పాల్గొంటున్నానని నాకు తెలుసు.

ఇది నా భావిశాయ సంరక్షణను ప్రభావితం చేయకుండా ఏ సమయంలోనైన నేను విరమించుకోనవచ్చునని నేను అర్థం చేసుకున్నాను.

ఈ పరిశోధనలో నా భాగస్వామ్యంతో నా గురించి సేకరించిన సమాచారం సంబంధిత వ్యక్తులు (నీటి కమిటి సభ్యులు/క్రమబద్ధీకరణ అధికారులు) చూడవచ్చునని నేను అర్థం చేసుకున్నాను. నా రికార్డులను ఈ వ్యక్తులకు అందుబాటులో ఉంచుటకు ఇష్టాను. నేను అర్థం చేసుకున్నది ఏమనగా ఈ అధ్యయనంలో కొత్త మందులు ఉపయోగించబడవు మరియు యధృద ప్రయోగశాల పరిక్షలు మాత్రమే నిర్వహించబడును.

_____                         _____
పేరు/వేలి ముద్ర                                       సంతకం/తేది

_____                         _____
పేరు/చట్టబద్ధంగా ఆమోదించబడే ప్రతినిధి (ఎల్ఎఆర్)          సంతకం/తేది

_____                         _____
తెలియపర్చబడిన అంగీకారం చర్చను నిర్వహించే వ్యక్తి సంతకం      సంతకం/తేది

_____                         _____
అంగీకారం పొందుచున్న వ్యక్తి పేరు                        సంతకం/తేది

सूचित सहमती प्रपत्र

इस अध्ययन के लिए विषय पहचान संख्या _____

परियोजना का शीर्षक

गंभीर ड्रग इंटरैक्शन के साथ-साथ दवाओं का आकलन जिसके परिणामस्वरूप हृदय रोग केयर अस्पताल में कार्डियक आउट पेशेंट में क्यूटी अंतराल लम्बाई का मूल्यांकन.

मार्गदर्शक (गाइड) का नाम: पि. सागर        टेलीफोन नंबर:8121898000

मुझे उपर्युक्त अध्ययन पर सूचना पत्र प्राप्त हुआ है और मैंने लिखित सूचना को पढ़ा और समझ लिया है.

मुझे अध्ययन पर चर्चा करने और प्रश्न पूछने का मौका दिया गया है. मैं समझता हूँ की अध्ययन के दौरान कोई नई दवा का उपयोग नहीं किया गायेगा और अध्ययन में केवल नियमित प्रयोगशाला परिक्षण शामिल है.

मैं अध्ययन में भाग लेने की सहमती देहा हूँ और मुझे पता है की मेरी भागीदारी स्वैच्छिक है.

मैं समझता हूँ की, मेरी भविष की देखभाल को प्रभावित किए बिना किसी भी समय मैं इस अध्ययन को वापस ले सकता हूँ.

मैं समझता हूँ की,

मेरे बारे में जानकारी एकत्र की जाती है जो इस अध्ययन मेमरी भागीदारी से ली जाती है और किसी भी मेडिकल not के अनुभाग जिम्मेदार व्यक्तियों (नैतिकता समिति के सदस्यों/नियामक प्राधिकरण) द्वारा देखे जा सकते है. मैं अपने रिकॉर्ड तक पहुँचने के लिए इन व्यक्तियों को पहुँच प्रदान करता हूँ. मैं समझता हूँ की इस अध्ययन में कोई नई दवा का उपयोग नहीं कया जा रहा है. इसके आलावा, अध्ययन में केवल नियमित प्रयोगशाला परीक्षायों के संचलन शामिल है.


_____                    _____
विषय का नाम और अंगूठे की छाप                         दिनांक/हस्ताक्षर


_____
कनोनी रूप से स्वीकृत प्रतिनिधि (LAR) का नाम


_____                    _____
सहमती प्राप्त करने वाले व्यक्ति का नाम                   दिनांक/हस्ताक्षर

نتائج سے اگا بی پر رضا مندی کا فارم

اس مطالعہ کے موضوع کا شناختی نمبر

ـــــــــــــــــــــــــــــــــــــــــــــ

پروجیکٹ کا عنوان:

سنگین منشیات کے منفی اثرات کا جائزہ لینے کے ساتھ ساتھ منشیات سے باہر مریضوں کے اندر QT کے وقفے کی طویل مدت میں منشیات کا تعین گیر ہسپتال پر

گائیڈ کا نام : پی۔ ساگر                    ٹیلی فون نمبر: 8121898000

میں نے مذکورہ بالا مطالعہ پر معلوماتی ورقہ حاصل کیا ہے اور تمام تحریری معلومات کو پڑھا/سمجھ چکا/چکی ہوں

مجھے مطالعہ کے بارے میں بات کرنے اور سوالات پوچھنے کا موقع فراہم کیا گیا، مجھے بتایا گیا ہے کے مطالعہ کے دوران کوئی جدید ادویات کا استعمال نہیں ہوگا، صرف معمول کے تجرباتی معینہ مطالعہ میں شامل ہوں۔

میں مطالعہ میں شرکت کے لئے رضا مندی دیتا/دیتی ہوں اور میں واقف ہوں کے میری شرکت اپنی مرضی سے ہے۔

مجھے بتایا گیا کی میں کسی بھی وقت دستبردار ہو سکتا/ہو سکتی ہوں جس سے میرا مستقبل علاج متاثر نہیں ہوگا۔

مجھے بتایا ہے کی اس مطالعہ میں شرکت سے میرے متعلق جمع کردہ معلومات اور کسی بھی میڈیکل نوٹس کے حصتوں کو زمیندار افراد (ایتھکس کمیٹی ممبران/ریگولیٹری اتھارٹی) کے جانب سے دیکھا جائنگا، میں ان افراد کو میرے ریکارڈز تک رسائی دیتا/دیتی ہوں۔ مجھے بتایا گیا ہے کی اس مطالعہ میں کوئی جدید ادویات کا استعمال نہیں کیا گیا۔ عالوازین، اس مطالعہ میں معمول کے لیبارٹری ٹسٹ شامل رہیں گے۔

تاریخ/دستخط                مریض کا نام/انگولھے کا نشان

تاریخ/دستخط                نام/خانودی ملور پر قبول نما8ندہ (LAR)

تاریخ/دستخط                رضا مندی کے حاصل کندہ کا نام

**CARE Hospital**
**Institutional Ethics Committee**
Regn. No. ECR/94/Inst./AP/2013/RR-16

Date: 11th Dec 2018

To

Afzoon Buttol and N. Shushrusha
Pharma-D Students,
Dept. of Electrophysiology,
Guru Nanak Institutions Technical Campus-School of Pharmacy,
Ibrahimpatnam, R. R. District, Telangana

Dear Madam,

Ref: Protocol – "Assessment of Serious Drug Interactions as well as Drugs Resulting in QT Interval Prolongation in a Cardiac Output at a Tertiary Care Hospital."

Sub: EC approval of the A14 of Institutional Ethics Committee Meeting held on 2nd Dec 2018

The Institutional Ethics Committee met on Sunday, 2nd Dec 2018 at 10.00 A.M. at Board Room (PACE), Care Convergence Centre, 8-2-595/2/B, Road No.10, Banjara Hills, Hyderabad–500034 for the monthly meeting and reviewed/discussed the thesis protocol submitted by you with the following documents of above mentioned study to be carried out at the site Dept. of Electrophysiology, Care Hospital, Hyderabad.

1. Letter from the student to the Chairman, Institutional Ethics Committee dated 15th Nov 2018
2. Protocol
3. Patient Information Sheet in English, Telugu, Hindi and Urdu
4. Informed Consent Form in English, Telugu, Hindi and Urdu

The following Members of Institutional Ethics Committee are present for the study

| Name | Qualification | Designation | Field |
|------|---------------|-------------|-------|
| Justice Ramakrishnam Raju | BA, BL | Chairman | Non Scientific/ Judiciary/ Legal Expert (Non Institutional) |
| Dr. B. V. Rama Raju | MS, Mch | Member Secretary | Scientific/ Medicine/ Clinician (Institutional) |

#8-2-595/2/B, Care Convergence Centre, Care Foundation, 2nd Floor,
Road No. 10, Banjara Hills, Hyderabad - 500 034, Telangana, INDIA
Phone: 91-40-39116069, Fax: 91-39116019, 23355316 | Email: ethicscommittee@carefoundation.org.in

Assessment of Drug Interactions & Drugs Resulting in QT Interval Prolongation